CELL BIOLOGY RESEARCH PROGRESS

BASOPHIL GRANULOCYTES

Cell Biology Research Progress

Tumor Necrosis Factor
Toma P. Rossard (Editor)
2009. ISBN: 978-1-60741-708-8

Tumor Necrosis Factor
Toma P. Rossard (Editor)
2009. ISBN: 978-1-61668-276-7 (Online book)

Cell Determination During Hematopoiesis
Geoffrey Brown and Rhodri Ceredig (Editors)
2009. ISBN: 978-1-60741-733-0

Handbook of Cell Proliferation
Andre P. Briggs and Jacob A. Coburn (Editors)
2009. ISBN: 978-1-60741-105-5

Handbook of Cell Proliferation
Andre P. Briggs and Jacob A. Coburn (Editors)
2009. ISBN: 978-1-60876-854-7 (Online book)

Daughter Cells of Microalgae
Dilwyn J. Griffiths
2010. ISBN: 978-1-60876-787-8

Handbook of Free Radicals: Formation, Types and Effects
Dimitri Kozyrev and Vasily Slutsky (Editors)
2010. ISBN: 978-1-60876-101-2

Endocytosis: Structural Components, Functions and Pathways
Brynn C. Dowler (Editor)
2010. ISBN: 978-1-61668-189-0

Endocytosis: Structural Components, Functions and Pathways
Brynn C. Dowler (Editor)
2010. ISBN: 978-1-61668-717-5 (Online book)

Cell Respiration and Cell Survival: Processes, Types and Effects
Gijsbert Osterhoudt and Jos Barhydt (Editors)
2010. ISBN: 978-1-60876-462-4

Handbook of Molecular Chaperones: Roles, Structures and Mechanisms
Piero Durante and Leandro Colucci (Editors)
2010. ISBN: 978-1-60876-366-5

Cell Division: Theory, Variants and Degradation
Yuri N. Golitsin and Mikhail C. Krylov (Editors)
2010. ISBN: 978-1-60876-986-5

Basophil Granulocytes
Paul K. Vellis (Editor)
2010. ISBN: 978-1-60741-797-2

Daughter Cells: Properties, Characteristics and Stem Cells
Ayane Hitomi and Masuyo Katoaka (Editors)
2010. ISBN: 978-1-60876-790-8

Cytoskeleton: Cell Movement, Cytokinesis and Organelles Organization
Sébastien Lansing and Tristan Rousseau (Editors)
2010. ISBN: 978-1-60876-559-1

Prostaglandins: Biochemistry, Functions, Types and Roles
Gillian M. Goodwin (Editor)
2010. ISBN: 978-1-61668-272-9

Prostaglandins: Biochemistry, Functions, Types and Roles
Gillian M. Goodwin (Editor)
2010. ISBN: 978-1-61668-645-1 (Online book)

Lipids: Categories, Biological Functions and Metabolism, Nutrition and Health
Paige L. Gilmore (Editor)
2010. ISBN: 978-1-61668-464-8

Lipids: Categories, Biological Functions and Metabolism, Nutrition and Health
Paige L. Gilmore (Editor)
2010. ISBN: 978-1-61668-522-5 (Online book)

CELL BIOLOGY RESEARCH PROGRESS

BASOPHIL GRANULOCYTES

PAUL K. VELLIS
EDITOR

Nova Science Publishers, Inc.
New York

LIBRARY OF CONGRESS CATALOGING-IN-PUBLICATION DATA
Basophil granulocytes / [edited by] Paul K. Vellis.
p. cm.
Includes index.
ISBN 978-1-60741-797-2 (hardcover)
1. Basophils. 2. Granulocytes. I. Vellis, Paul K.
QP95.2.B37 2009
571.9'68--dc22
2009052727

Published by Nova Science Publishers, Inc. ✢ New York

Contents

PREFACE

This new book examines the latest research developments in the field of basophil granulocytes, sometimes referred to as basophils, which are the least common of the granulocytes, representing about 0.01% to 0.3% of circulating white blood cells. The name comes from the fact that these leucocytes are basophilic, i.e., they are susceptible to staining by base dyes, as shown in the picture. Basophils contain large cytoplasmic granules which obscure the cell nucleus under the microscope. However, when unstained, the nucleus is visible and it usually has 2 lobes. The mast cell, a cell in tissues, has many similar characteristics.

Chapter 1- Basophils and mast cells have long been recognized as critical effector cells in IgE-mediated immune responses.

Antigen interaction with specific IgE bound to the high-affinity Fc receptor for IgE, constitutively expressed on the cell-surface of basophils, generates signals that cause a shift in the resting state equilibrium of phosphorylation and dephosphorylation events that serves to maintain homeostasis. The outcome of this activated state is the release of a wide array of pro-inflammatory mediators.

During the past few years, the existence of a negative feedback loop initiated upon FcεRI engagement has also been envisaged. This negative signal involves the coordinated action of adaptors, phosphatases and ubiquitin ligases that limits the intensity and duration of positive signals, thus modulating basophil functions. Relevant to this, others and we have demonstrated that Cbl proteins control the amplitude of FcεRI-generated signals mainly by specific ubiquitin modification of activated receptor subunits and associated protein tyrosine kinases.

The identification of molecular mechanisms involved in basophil regulation, may provide new insights into how these cells can be manipulated to achieve therapeutic ends in the treatment of allergic disease.

Chapter 2- Liver cancer is still a major public health problem worldwide. It is now known that solid tumors, including hepatocellular carcinoma, are infiltrated by immune cells most prominently T and B lymphocytes. All of them are variably scattered within the tumor and loaded with an assorted array of cytokines, chemokines, and inflammatory and cytotoxic mediators. This complex network reflects the diversity in tumor biology and tumor-host interactions.

Basophil granulocytes and mast cells are well established effector cells in IgE-associated immune responses. While in allergies or parasitic infections the role of these cells has been known for years, in cancer there are what seem to be conflicting data describing a supporting or a possibly inhibitory role of these cells in several human neoplasia.

Here, these knowledge will summarize, focusing on the implications of these findings in the understanding the role of basophil granulocytes and mast cells in primary and secondary liver cancer.

Chapter 3- Basophilic granulocytes (basophils) are a very small population of peripheral blood leukocytes. Because basophils have high-affinity immunoglobulin E receptors (FcεRI) and secrete chemical mediators that contain histamine, they are thought to be very similar to mast cells. However, research characterizing the function of basophils was slow to proceed and their unique function and importance were not established for a significant period of time. Recently, a series of studies have characterized the role of basophils in anaphylactic shock, chronic allergic reactions, and other human immunological reactions. These studies have shown that basophils are not a supplementary cell type but key players in very serious immune reactions. In this chapter we introduce the recently discovered characteristics of basophils. In addition, we consider how these aspects are clinically important and are connected with new cellular or molecular treatments for allergic reactions.

Chapter 4- Basophils comprise less than 1% of blood circulating granulocytes and share several similarities with mast cells, including expression of high affinity IgE receptor and the ability to secrete chemical mediators and cytokines. Basophils also express CD40 ligand, which in combination with IL-4 and IL-13 promotes IgE class switching in B cells. Flavonoids, ubiquitously present in vegetables, fruits or teas possess inhibitory activity of basophil activation. Flavonoids inhibit histamine release, synthesis of IL-4 and IL-13 and CD40 ligand expression by activated basophils. Among flavones, luteolin, apigenin and fisetin are the strongest inhibitors of IL-4 production with an IC_{50} value of 2-6 □M, while quercetin and kaempferol exert a substantial inhibitory activity with an IC_{50} value of 15-19 □M. The inhibitory activity of flavonoids against IL-4 and CD40 ligand expression is thought to be mediated through their inhibition action of activation of the nuclear factor of activated T cells and AP-1. These suppressive effects of flavonoids on basophil activation suggest that an appropriate daily intake of flavonoids might ameliorate allergic symptoms or prevent the onset of allergic diseases. In fact, administration of astragalin, a kaempferol glucoside, to atopic dermatitis-prone mice was found to prevent the onset of dermatitis and ameliorated the severity of dermatitis even after onset. In addition, intake of enzymatically modified isoquercitrin, a quercetin glycoside, had an ameliorative effect on ocular symptoms of patients with Japanese cedar pollinosis.

Chapter 5- Basophils are a minute group of granulocytes. Among types of white blood cells, they are the least in number. Knowledge on the basophil is limited. In this article, the author will briefly detail and discuss basophils in relation to tropical infections. The scenarios of malaria, dengue, tuberculosis as well as leprosy will be presented.

Chapter 6- Until now, the basophil had been among the least studied types of white blood cells. Basically, medical scientists know that the main role of the basophil relates to an anaphylaxis reaction. However, there are many other facets of basophil roles. In this chapter, the author will briefly summarize important reports on the basophil in Thailand.

Chapter 7- Basophils are the least abundant granulocytes in the blood, barely reaching 1% of the total leukocyte population. These cells have long been ignored and their function in health and disease has remained an enigma [1]. In fact, basophils have been viewed by many as "redundant circulating mast cells". Recent advances, however, towards understanding the development of both mast cells and basophils, has allowed identification of key factors leading to the production of these cell types [2]. In humans, it is known that basophils are derived from $CD34^{+}$ bone marrow hematopoietic progenitor cells that yield $Fc\varepsilon RI\alpha^{hi}c\text{-}kit^{-}$ cells [2]. In the mouse, a common spleen progenitor leading to basophils and mast cells has been described [2]. Basophils arise from this GATA2-expressing progenitor by up regulation of the transcription factor C/EBPα. In contrast, downregulation of C/EBPα together with expression of GATA2 leads this common precursor to generate mast cells [3]. However, whether these common spleen precursors lead to both kinds of cells *in vivo* remains to be established.

The lack of a basophil-deficient mouse model as well as the need for a differentiated basophil cell line has hampered studies on the role of basophils in health and diseases. Indeed, until the recent advance allowing antibody-mediated basophil depletion in mice and identification of a panel of surface marker for mouse basophils [4], the study of these cells has been limited to human patho-physiological approaches. The latter allowed the identification of basophils as important players *in vivo* in pathologies like asthma, allergic diseases and parasitic infections [1]. Like mast cells, basophils express the tetrameric form ($\alpha\beta\gamma_2$) of the high affinity receptor for immunoglobulin E (IgE) FcεRI [5]. Basophils also express a variety of other surface receptors that can lead to their activation (e.g. CD123, TLR4, FcγRIII…) [6]. FcεRI-mediated activation of basophils induces, as in mast cells, the immediate release of preformed pro-inflammatory mediators (such as histamine and proteases) and the later production and secretion of leukotrienes, cytokines and chemokines (such as IL-8, IL-4, IL-5, IL-13 and IL-10). These responses (both immediate and delayed) are potentiated with IL-3 pre-incubation of the cells [1,6].

Chapter 8- Described by Paul Ehrlich in 1879, basophilic granulocytes (basophils) remain among the least studied white blood cells. A rather small number of works on them (compared to other leukocyte types) was attributable to their extremely low content in the blood (0,5-1% of nucleated blood cells), but not to lack of interest by scientists. The number and morphology of basophils are somewhat different in different animals, but they occur in all classes of vertebrates. It demonstrates the importance of these cells for protective systems. Basophils are involved in allergic reactions and anti-parasitic protection. Moreover, due to cytokine secretion, in particular IL-4 and IL-13, they became the focus of attention as a cell type implicated in the immune response switch toward Th2 [Falcone et al., 2006]. The relationship of basophils with the acute phase of inflammation factors, in particular, with C-reactive protein (CRP), are poorly elucidated.

Elevated expression of CRP, a pentraxin and known marker of acute phase of inflammation, accompanies the onset of any form of immune response. However, CRP influence on the intensity and specificity of immune reactions is not clearly understood. It has been known that a good correlation exists between CRP production at early stage of immunization and subsequent antibody production in rabbits. It has been also shown that CRP could influence the fine specificity of immune responses to antigens which are its ligands. CRP could decrease protective antibody production against pneumococcal cell wall phosphorylcholine (PC). PC binding by CRP leads to the opsonization of pneumococcci, complement and phagocytosis activation [Nakayama S. et al., 1984; Horowitz J. et al., 1987; Szalai A.J. et al., 1995]. There are numerous data concerning stimulating effects of CRP on phagocytic activity and other functions of macrophages and neutrophils. CRP has been shown to depend for its biological function on cellular Fcγ receptors (FcgRs). FcgRI is a high-affinity IgG receptor expressed on myeloid cells that is up-regulated during inflammation [Van de Winkel J.G.J.,

Anderson,C.L., 1991; Van de Winkel J.G.J., Capel P.J., 1993; Beekman J.M. et al., 2004]. However, the major receptor for CRP on leukocytes is low affinity FcgRII [Bharadwaj D. et al., 1999].

In: Basophil Granulocytes
Editor: Paul K. Vellis

ISBN: 978-1-60741-797-2

Chapter 1

NEGATIVE REGULATION OF FCεRI-MEDIATED BASOPHIL ACTIVATION BY THE CBL FAMILY OF ADAPTOR PROTEINS

Rosa Molfetta, Francesca Gasparrini, Angela Santoni and Rossella Paolini
Department of Experimental Medicine, Sapienza University, 00161 Rome, Italy.

ABSTRACT

Basophils and mast cells have long been recognized as critical effector cells in IgE-mediated immune responses.

Antigen interaction with specific IgE bound to the high-affinity Fc receptor for IgE, constitutively expressed on the cell-surface of basophils, generates signals that cause a shift in the resting state equilibrium of phosphorylation and dephosphorylation events that serves to maintain homeostasis. The outcome of this activated state is the release of a wide array of pro-inflammatory mediators.

During the past few years, the existence of a negative feedback loop initiated upon FcεRI engagement has also been envisaged. This negative signal involves the coordinated action of adaptors, phosphatases and ubiquitin ligases that limits the intensity and duration of positive signals, thus modulating basophil functions. Relevant to this, others and we have demonstrated that Cbl proteins control the amplitude of FcεRI-generated signals mainly by specific ubiquitin modification of activated receptor subunits and associated protein tyrosine kinases.

The identification of molecular mechanisms involved in basophil regulation, may provide new insights into how these cells can be manipulated to achieve therapeutic ends in the treatment of allergic disease.

INTRODUCTION

Although basophils account for less than 1% of peripheral blood human and murine leukocytes, they have long been recognized, together with tissue-resident mast cells, as the critical effector cells in IgE-mediated allergic diseases [1,2].

Basophils and mast cells share the presence of basophilic granules in their cytoplasm and the surface expression of a high affinity receptor for the Fc fragment of IgE (FcεRI).

FcεRI belongs to the multisubunit immunoreceptor family that lacks intrinsic enzymatic activity but transduces intracellular signals through association with cytoplasmic protein tyrosine kinases (PTKs) [3-5]. In rodent and human basophils, the intracellular signalling generated upon engagement of receptor-bound IgE with the corresponding allergens are responsible for the release of preformed and newly synthesized mediators including histamine, leukotrienes, IL-4 and IL-13 [6-8].

Besides these positive signals, FcεRI aggregation has been understood to generate negative intracellular signals capable of limiting basophil and mast cell functional responses through the action of a variety of multidomain adaptor proteins [9].

Among them, the Cbl family of ubiquitin (Ub) ligases [10-13] has attracted considerable interest due to the finding that it controls the intensity and duration of FcεRI-mediated signals mainly by specific Ub modification of the activated receptor subunits and associated PTKs [14-16].

Cbl could also promote internalization of engaged FcεRI through a pathway that is functionally separable from its Ub ligase activity and is dependent on Cbl interaction with a multidomain protein mainly involved in the process of endocytosis and vesicle trafficking, namely CIN85 (Cbl-interacting protein of 85 kDa) [17].

This chapter is aimed at providing an overview on the mechanisms through which Cbl proteins negatively regulate basophil functional responses.

The High Affinity Receptor for IgE: Structure and Function.

FcεRI is expressed on mast cells and basophils as a heterotetramer composed by an IgE-binding α subunit, a four transmembrane-spanning β subunit, and two identical disulphide-linked γ subunits [3,4]. In rodents, all three subunits are needed for surface expression, whereas in humans a trimeric complex lacking the β-chain (FcεRI$\alpha\gamma_2$) also exists, and is expressed not only on mast cells and basophils but also on antigen presenting cells [7].

FcεRIβ subunit, whose levels are increased in human basophils cultured in the presence of IL-3 [18], acts as an amplifier of FcεRI surface expression by controlling the proper trafficking and maturation of the α-chain [19].

FcεRI surface expression is also positively regulated by the presence of circulating IgE [20]. The level of serum IgE correlates with the cell surface density of FcεRI on human basophils [21-23], and monomeric IgE upregulates the surface expression of FcεRI by a direct interaction with the receptor itself [24-26]. Conversely, the treatment of atopic patients with anti-IgE antibodies decreases serum IgE levels and FcεRI surface expression [27].

Several observations obtained using a rat basophilic leukemia cell line, namely RBL-2H3, have contributed to elucidate the function of the different receptor subunits.

FcεRI α-chain contains two extracellular Ig-like domains involved in IgE binding, a transmembrane region with an aspartic acid residue and a short cytoplasmic tail that lacks signal transduction motifs. The β and γ subunits have no role in ligand binding, but they share a conserved immunoreceptor tyrosine-based activation motif (ITAM) within their long cytoplasmic tails that, upon FcεRI aggregation, is rapidly phosphorylated on tyrosines by the Src family kinase Lyn that binds to FcεRIβ under resting conditions [5,7].

The phosphorylated γ-chains are competent to drive cell activation in the absence of a β-chain ITAM, however FcεRIβ amplifies the intensity of signals from FcεRIγ [28], thus the two subunits act cooperatively in promoting signal transduction.

It has been demonstrated that the full activation of FcεRI requires its migration into lipid rafts, specialized regions of the plasma membrane enriched in cholesterol and glycosphingolipid that form ordered but dynamic structures floating in the less ordered surrounding membrane [29]. Upon FcεRI engagement membrane rafts coalesce into larger and more stable structures where engaged receptors are concentrated [30], and can more easily interact with signalling molecules, such as active Lyn [31,32], thus favouring ITAM phosphorylation.

Phosphorylated ITAMs provides a docking site for the tandem pair of Src homology 2 (SH2) domains of the cytoplasmic kinase Syk, that is, in turn, activated upon tyrosine phosphorylation [5]. The use of Syk specific inhibitors and Syk-negative RBL-2H3 cells has demonstrated an obligatory role for this kinase in FcεRI-mediated signalling [33-36]. In humans, a minority of normal blood donors contains basophils that fail to degranulate, and these "nonreleaser" basophils express normal level of FcεRI but very low levels of Syk protein [37,38]. Furthermore, variable expression levels of Syk, observed in human basophils from different releaser donors, correlate well with the IgE-mediated responsiveness of these cells [39].

Syk activation enables the productive interaction with its many targets including the membrane anchored linker for activation of T cells (LAT) that, once phosphorylated, recruits SH2-containing adaptors such as leukocyte protein of 76kDa (SLP-76) and Grb2, and enzymes such as phospholipase Cγ (PLCγ). Once in the membrane, PLCγ undergoes activating tyrosine phosphorylation, and hydrolyzes the membrane phosphatidyl inositol 4,5-bisphosphate [PtdIns(4,5)P_2] to form the soluble inositol 1,4,5-triphosphate (IP_3) and the membrane bound diacylglycerol (DAG), which are responsible for intracellular calcium mobilization and protein kinase C (PKC) activation, respectively.

The adaptors SLP-76 and Grb2 recruit exchange factors promoting the activation of the small GTPases, Ras, Rac, and Rho. They regulate complex networks of signalling pathways leading to the secretion of preformed and newly synthesized mediators and cytokines.

In RBL-2H3 cells, a complementary pathway is initiated by another PTK of the Src family, Fyn, and cooperates with the Lyn/Syk pathway in the propagation of FcεRI-mediated signal [40]. Upon receptor engagement, Fyn phosphorylates the molecular adaptor Gab2, favouring membrane recruitment of the p85 regulatory subunit of phosphatidylinositol-3-OH kinase (PI3K). Once activated, PI3K catalyzes the formation of PtdIns(3,4,5)P_3 (PIP_3) that functions as docking site for pleckstrin homology (PH) domain-containing proteins such as Bruton's tyrosine kinase (Btk) and PLCγ.

It is unclear whether the Fyn pathway controls PI3K activation in human basophils [8]. Regardless, the PI3K pathway plays an essential role in governing the production and release of all mediator classes in basophils, as demonstrated by the use of specific PI3K inhibitors that affect degranulation as well as generation of arachidonic acid metabolites and cytokine secretion [41,42].

Mechanisms Underlying Negative Regulation of Basophil Activation.

Over the past several years, it has become apparent that basophils express surface receptors that counteract FcεRI-mediated activation responses. The common feature of these inhibitory receptors is the presence in their cytoplasmic tail of immunoreceptor tyrosine-based inhibitory motifs (ITIM) then, once phosphorylated, recruits negative signaling molecules [43,44].

In human basophils, coaggregation of FcεRI with the ITIM containing FcγRIIB induces the recruitment and activation of the tyrosine phosphatase SHP-1 to the phosphorylated ITIM with consequent inhibition of FcεRI-mediated cell activation [45]. Thus, mechanistically, ITIM-bearing receptors suppress FcεRI-mediated signaling by promoting dephosphorylation events.

Furthermore, FcεRI signaling itself has been understood to consist of a mixture of positive and negative signals whose integration determines the rate and the extent of functional basophil responses. In RBL-2H3 cells, the phosphatases SHP-1, SHP-2 and SHIP are tyrosine phosphorylated and activated upon their recruitment to the phosphorylated FcεRIβ, thus promoting dephosphorylation events that contrast FcεRI-mediated signal propagation [46,47]. A low level of SHIP was detected in "hyper-releasable" basophils derived from highly allergic donors, and correlated with a higher sensitivity to stimulation [48], suggesting a role for SHIP as negative regulator of human basophil degranulation as well. Relevant to this, a negative correlation between maximum SHIP phosphorylation and basophil releasability was also observed [39].

Interestingly, negative signals can also be generated by signaling molecules, including PTKs and adaptors, initially thought to contribute only activating signals [9].

The concomitant discovery of increasing numbers of new inhibitory adaptors and their molecular interactions has revealed an intricate balancing system that contributes significantly to regulate basophil functions. Among them, Cbl family proteins have emerged as negative regulators of FcεRI-mediated signals.

Cbl Family Proteins

Domain Structure and Function of Cbl Proteins.

The mammalian Cbl family consists of three proteins encoded by separate genes: c-Cbl, Cbl-b and Cbl-3 [10-13] (Figure 1). All of them are structurally characterized by the presence of highly conserved regions in their N-terminal: a tyrosine kinase-binding (TKB) domain, a RING finger domain, and a proline-rich domain. Moreover, c-Cbl and Cbl-b share additional regions in their C-terminal mainly involved in protein-protein interactions [49]: a more extensive prolin-rich domain able to interact with several SH3-containing proteins, such as Src family kinases and CIN85/CMS family of adaptor proteins; a region containing several tyrosine residues which are phosphorylated following the stimulation of a diverse array of membrane receptors and promotes interaction with SH2-domain containing proteins; a region homologous to both Ub-Associated (UBA) domain and to Leucine Zipper (LZ) that mediate Ub binding and intermolecular oligomerization.

Importantly, the highly conserved TKB and RING finger domains define the basic functional unit of c-Cbl and Cbl-b [13, 49]. The TKB domain is composed of a four-helix bundle (4H), a calcium-binding EF hand and a modified SH2 domain, and was so named for its ability to bind to phosphotyrosine residues of receptor and non-receptor PTKs including growth factor receptors, such as EGFR and PDGFR, and cytoplasmic tyrosine kinases of the Syk/ZAP-70 family [50].

The RING Finger domain, interacting with Ub-conjugating enzymes, catalyzes protein ubiquitination [51,52].

Ubiquitination is a post-translational reversible modification whereby Ub, a 76-amino-acid-globular peptide (8 KDa), is covalently attached to lysine residues of acceptor proteins that are then mainly targeted to degradation [53-55]. Ubiquitination is catalyzed by the action of three different enzymes, namely E1, E2 and E3 (Figure 2A). The Ub-activating enzyme (E1) forms a thiol-ester bond with the carboxy-terminal glycine of Ub in an ATP-dependent process. Activated Ub is, successively, accepted by the Ub-conjugating enzyme (E2) by transthiolation, and finally transferred to the substrate through the action of the Ub protein ligase (E3). Thus, this latter class of enzymes provides specificity to the Ub system being responsible for substrate recognition and Ub ligation to the target protein.

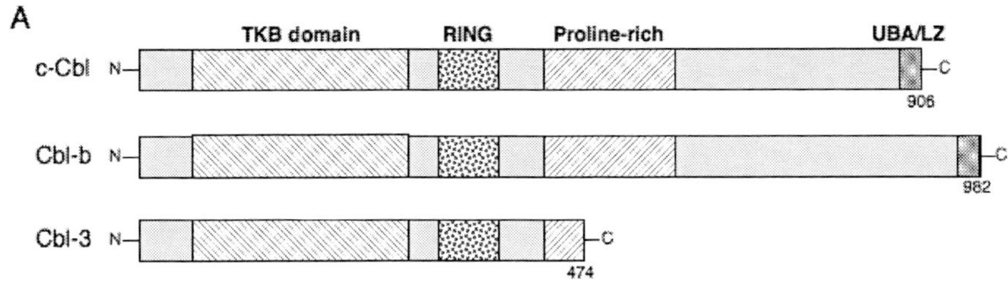

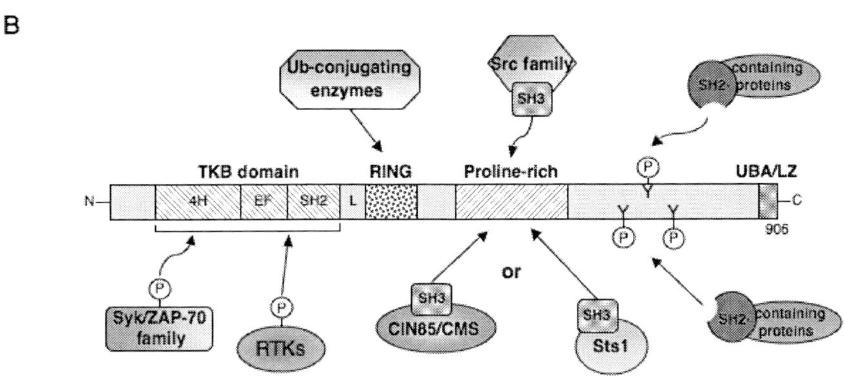

Figure 1 A. Schematic representation of the structural domains of the three mammalian Cbl (Casitas B-lineage lymphoma) isoforms c-Cbl, Cbl-b, and Cbl-3. B. The domains of c-Cbl and Cbl-b and their interactions with signalling proteins are outlined. The tyrosine-kinase-binding (TKB) domain mediates binding to phosphorylated tyrosines on tyrosine kinases. The RING-finger domain recruits ubiquitin-conjugating enzymes. The region between the TKB and the RING finger is referred to as the linker region (L) and is often deleted or mutated in oncogenic Cbl variants. The C-terminal proline-rich region is involved in the adaptor functions of Cbl proteins, by mediating binding of several Src-homology-3 (SH3)-containing proteins. The C-terminal UBA/LZ domain has a role in Cbl oligomerization and ubiquitin binding. Note that some of the phosphotyrosine residues, which control the binding with SH2-containing proteins, are absent from Cbl-3. CIN, c-Cbl-interacting protein; CMS, Cas ligand with multiple SH3 domains; RTKs, receptor tyrosine kinases; Sts, Suppressor of T-cell receptor Signalling. Modified from *ref. 10*, Thein and Langdon 2001.

The Cbl family belongs to the E3 Ub ligases characterized by the presence of a RING finger domain [52]: TKB domain determines Cbl substrate specificity serving as a docking site for tyrosine phosphorylated proteins that are then ubiquitinated by the RING-finger associated E2 enzyme (Figure 2B).

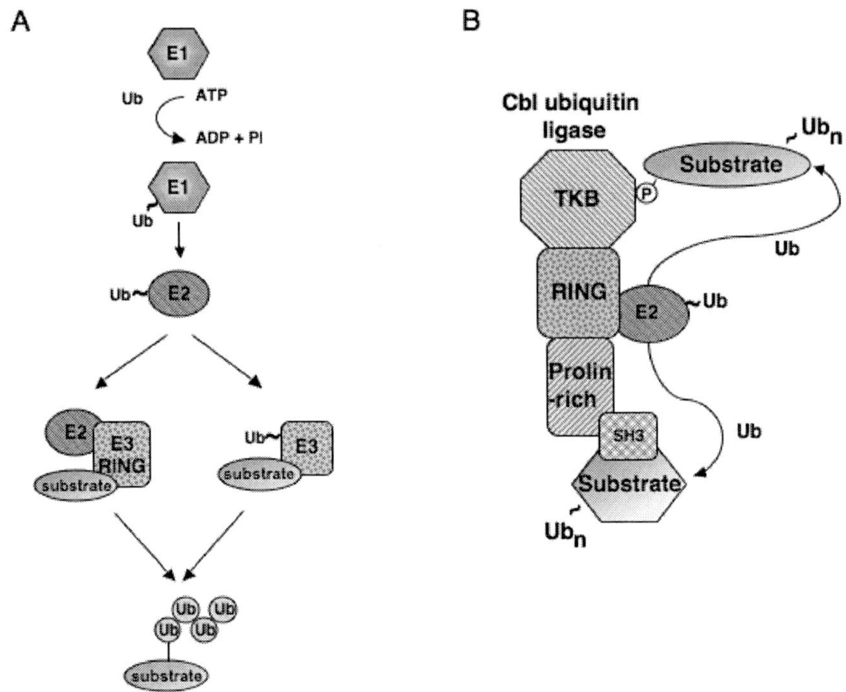

Figure 2 A. Schematic representation of the ubiquitination process. A hierarchical set of three types of enzyme is required for substrate ubiquitination: ubiquitin-activating (E1), ubiquitin-conjugating (E2) and ubiquitin-protein ligase (E3) enzymes. The two major classes of E3 ligases are depicted. Modified from Woelk et al., Cell Division, 2007. B. Model of Cbl Ubiquitin Ligase Function. Multiple motifs in Cbl proteins, such as the TKB domain, or the proline-rich domain, serve to recruit the substrates for ubiquitination. The ubiquitin-conjugating enzyme (E2), which interacts with the RING domain of the ubiquitin ligase Cbl (E3), transfers Ub to the target proteins. Modified from Duan L. et al, Immunity, 2004.

The role of substrate recognition is not carried exclusively by the TKB domain, since some substrates, such as Src family kinases, can be targeted by the additional domains located in the C-terminal region of both c-Cbl and Cbl-b.

Another factor essential for the E3 activity of c-Cbl (and in a lesser extent of Cbl-b) is the tyrosine phosphorylation of its linker domain located between the TKB and RING domains. This phosphorylation causes conformational changes in c-Cbl that are required for the release of the RING-finger associated E2, an essential step in Cbl driven ubiquitination [56].

Ubiquitinated substrates are, then, targeted to different degradation pathways depending on the kind of ubiquitination itself. Polyubiquitin chains, whereby Ub acts as a substrate for the attachment of further Ub molecules, is commonly recognized as a signal that targets substrates to degradation by the 26S proteasome [53-55, 57]. The attachment of single Ub to one or more lysines of target proteins, namely mono and multiubiquitination respectively, can act as internalization signals driving transport of membrane receptors along the endocytic pathway towards a lysosomal compartment for degradation [58-60].

In vitro, c-Cbl and Cbl-b appear to have equal capacity to act as E3 Ub ligases toward a similar range of substrates. However, unlike c-Cbl, ubiquitination by Cbl-b does not often result in substrate degradation, but rather appears to affect protein localization [49].

FcεRI Down-Regulation by Cbl-Mediated Ubiquitination.

Several reports have firmly established that recruitment of Cbl family proteins to activated RTKs mediates receptor ubiquitination and their sorting to lysosomes for degradation [59-63].

Evidence collected in the past years has strongly supported a similar crucial role for Cbl family Ub ligases in the down-regulation of immunoreceptors, including FcεRI [14,64,65].

Both rodent and human basophils express c-Cbl, which become tyrosine phosphorylated upon FcεRI engagement [66,67].

The first evidence for a negative role played by c-Cbl in basophils came from experiments in which c-Cbl overexpression inhibited receptor-mediated serotonin release in RBL-2H3 cells without affecting receptor phosphorylation [68]. The molecular mechanism underlying this inhibition has been elucidated by additional works demonstrating the involvement of c-Cbl in the ubiquitination of FcεRI receptor subunits. An earlier study from our group [69] had demonstrated that FcεRI β and γ subunits were subjected to ubiquitination upon stimulation of RBL-2H3 cells with IgE and multivalent antigen. Subcellular fractionation and confocal microscopy experiments have subsequently demonstrated that c-Cbl colocalizes with FcεRI β and γ subunits into lipid rafts after receptor engagement, suggesting the involvement of Cbl in receptor ubiquitination [70].

Our group have identified c-Cbl as the main E3 ligase responsible for the antigen-induced receptor ubiquitination in RBL-2H3 cells [14]. Overexpression of wild type c-Cbl but not a mutant form deleted in the RING finger domain increased antigen-induced FcεRI β and γ ubiquitination, providing evidence for a direct role of c-Cbl as Ub ligase. Furthermore, proteasome inhibition induced a persistence of tyrosine phosphorylated β chains, supporting the involvement of Cbl-dependent ubiquitination in the down-regulation of engaged receptors.

In line with this finding, we have more recently demonstrated that FcεRI β and γ subunits are mainly monoubiquitinated by c-Cbl at multiple sites upon Ag stimulation, and provided evidence that this modification controls receptor internalization and sorting along the endocytic compartments through the interaction with Ub-binding adaptors [71]. In particular, we have demonstrated a critical role for Hrs in controlling the fate of internalized receptor complexes: Hrs depletion retains ubiquitinated receptors into early endosomes and partially prevents their sorting into lysosomes. Our results are consistent with previous studies showing that Hrs depletion induces retention of ubiquitinated EGFR in early endosomes, and impairs degradation of internalized receptors [72,73].

Since FcεRI β and γ subunits and c-Cbl have been reported to translocate into lipid rafts upon Ag stimulation [30,31,70], we have also investigated the implication of a lipid raft environment in regulating Cbl-mediated FcεRI ubiquitination. We demonstrate that the recruitment of engaged FcεRI subunits into lipid rafts precedes their ubiquitination, and that the integrity of these membrane microdomains is required to allow receptor ubiquitination [71]. We also show a strong interdependence between lipid rafts and receptor endocytosis, in line with the finding of Fattakhova and coworkers, who demonstrated that aggregated FcεRI complexes remain associated with lipid rafts upon Ag-induced internalization [74].

All together our data demonstrate that Cbl-dependent FcεRI ubiquitination initiates into lipid raft and provide evidence that FcεRI may use Ub as an internalization signal. Furthermore, they support a key role for the Ub pathway to ensure proper endocytic trafficking of an immune receptor to the lysosomal compartment where degradation of the complexes can take place (Figure 3).

FcεRI Down-Regulation by Cbl Adaptor Function.

Some evidence indicate that Cbl proteins can promote RTK down-modulation independently by their ligase activity through the interaction with the multidomain protein CIN85 (Cbl-interacting protein of 85 kDa) that is constitutively associated to endophilin, a regulatory component of clathrin-coated pit formation [75,76].

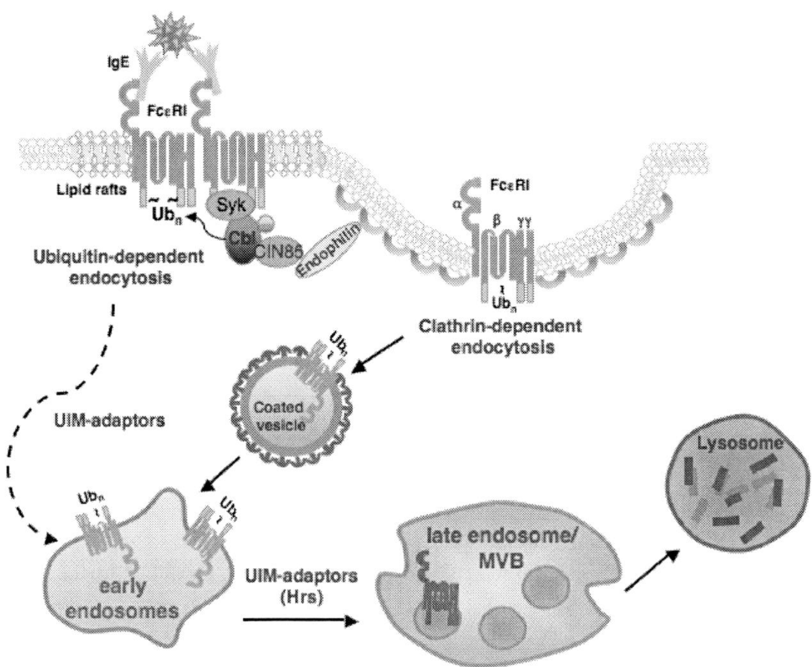

Figure 3. Schematic model of ligand-induced FcεRI endocytosis by Cbl ubiquitin ligase and by Cbl adaptor function. Upon FcεRI engagement, Cbl promotes receptor ubiquitination and the recruitment of CIN85/endophilin in the proximity of engaged receptor complexes, whereby endophilins drive their clathrin-mediated endocytosis. Furthemore, receptor multiubiquitination may act as an independent signal for receptor endocytosis. Internalized receptor complexes are then sorted along the endocytic pathway and targeted to the multivesicular bodies and lysosomes for degradation.

CIN85 is a broadly expressed adaptor protein characterized by the presence of three SH3 domains and a proline-rich region, both involved in protein-protein interaction [77].

CIN85 binding to Cbl is mediated by its SH3 domains and is enhanced by ligand-induced tyrosine phosphorylation of Cbl, while the proline-rich region of CIN85 constitutively interacts with endophilins.

In RBL-2H3 cells, we found a Cbl/CIN85 interaction that increases upon antigen driven FcεRI stimulation [17]. Moreover, CIN85 overexpression results in a more rapid receptor internalization, entry into early endosomes, and delivery to a lysosomal compartment for degradation. In the same study we have also demonstrated an impairment of FcεRI-mediated degranulation in CIN85 transfectants, suggesting that the accelerated FcεRI internalization may contribute to dampen cell functional responses.

In conclusion, our data strongly favour a model in which after receptor engagement the Cbl/CIN85 complex is recruited to the cell membrane where it can drive internalization of engaged FcεRI complexes and their sorting into endocytic compartments, a process required for receptor degradation (Figure 3).

Cbl-Mediated Down-Regulation of Non-Receptor PTKs Upon FcεRI Engagement.

Cbl proteins induce ubiquitination of a number of non-receptor PTKs, including Syk and several members of Src family kinases, and in most case ubiquitination of activated PTKs correlates with downregulation of their protein level and kinase activity [14,64,67,78-82].

In RBL-2H3 we have demonstrated that, following FcεRI stimulation, c-Cbl is responsible for Syk ubiquitination, and that Syk activity controls its own ubiquitination [14] (Figure 4).

Syk and c-Cbl have been previously reported to be constitutively associated in RBL-2H3 cells [66]. Thus, it is conceivable that Syk/Cbl interaction allows the enzymes to become reciprocal substrates: Syk phosphorylates and activates Cbl that, in turn, ubiquitinates activated Syk.

In line with this model, we have demonstrated that ubiquitination preferentially affects the phosphorylated and active forms of Syk [14]. Moreover, proteasome inhibition induced a persistence of activated kinase forms, supporting the involvement of Cbl-dependent ubiquitination in the down-regulation of the active pool of Syk.

In human basophils FcεRI stimulation, achieved by antigen or anti-IgE antibodies, promotes a progressive loss of Syk protein that correlates with Syk ubiquitination and is sensitive to proteasome inhibitors [67]. Moreover, an antigen-dependent interaction between Syk and Cbl was observed, and the level of protein association correlates with that of Cbl tyrosine phosphorylation, suggesting a role for Cbl in targeting Syk ubiquitination and degradation.

This hypothesis has been supported by more recent evidence showing that also IgE-independent stimuli are able to induce c-Cbl phophorylation, and are responsible for Syk degradation and for the induction of nonreleasing phenotype [83].

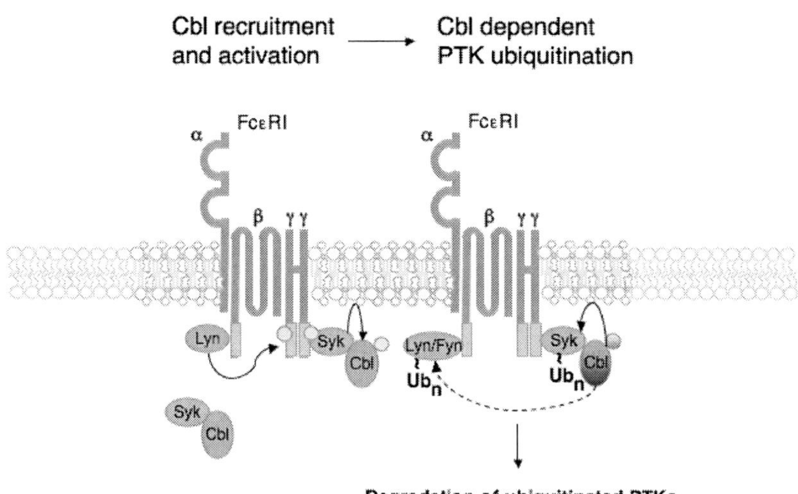

Figure 4. Model of proten tyrosine kinase ubiquitination by Cbl. Upon Lyn-mediated receptor phosphorylation, a preformed Syk/Cbl complex binds to phosphorylated FcεRI subunits. Syk phosphorylates and activates Cbl that in turn ubiquitinates Syk and Src family protein tyrosine kinases.

Following FcεRI engagement, Cbl proteins can also ubiquitinate Lyn and Fyn in RBL-2H3 cells [15]. Both c-Cbl and Cbl-b act as Ub ligases targeting the active forms of Lyn, further supporting our previous finding that ubiquitination preferentially affects active forms of the kinases responsible for signal propagation. The same study also provided evidence for the requirement of a lipid raft environment in Cbl-mediated ubiquitination events: targeting Cbl-b into lipid raft enhances Lyn ubiquitination. Moreover, overexpression of a mutant form of Cbl-b that constitutively localizes in lipid rafts reduces antigen-mediated degranulation and cytokine production by negatively regulating both Lyn-Syk-LAT and Fyn-Gab2-mediated signalling pathways [16]. At molecular level, the membrane-targeted overexpression of Cbl-b inhibited receptor phosphorylation and kinase activity, and dramatically downregulated the protein amount of Gab2 by promoting its ubiquitination.

Thus, overexpression studies in RBL-2H3 cells have shown that c-Cbl and Cbl-b play similar negative regulatory roles in FcεRI-mediated signaling. Additional studies are needed to achieve a similar conclusion also for human basophils.

Cbl-Mediated Regulation of PTKs in Unstimulated Basophils.

Evidence collected in the past years have demonstrated that human basophils show a wide variability in the rate of FcεRI-mediated degranulation and, among them, a small percentage of basophils completely fail to release histamine after FcεRI crosslinking. Their unresponsiveness has been linked to lower levels of Lyn and to the absence of Syk despite a normal level of their mRNA, suggesting the action of a post-translational mechanism responsible for kinase degradation [37,38]. This mechanism appears to be cell specific, since B lymphocytes, eosinophils, and neutrophils purified from the same donors have normal Syk and Lyn levels, and is reversible since the donors can fluctuate between the nonreleaser and the releaser phenotypes with a concomitant change of Syk expression [37], and because unresponsiveness can be overcome by IL-3 treatment [84].

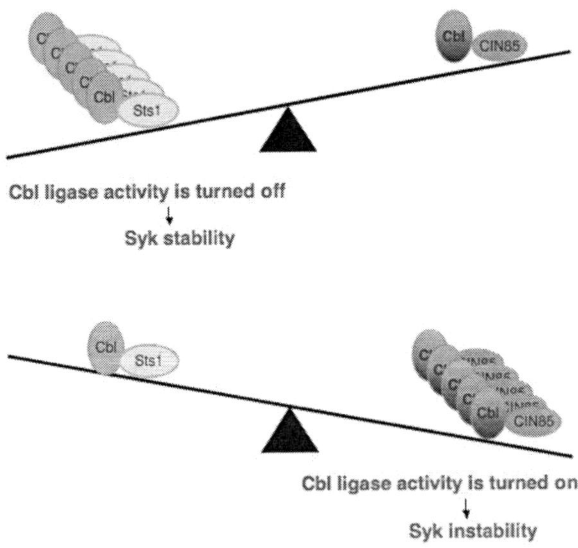

Figure 5. Schematic model of CIN85 and Sts1 antagonist action on Cbl ligase activity. Sts1 and CIN85 compete to bind Cbl and exert an antagonist action on its ligase activity: in RBL-2H3 cells Cbl binding with Sts1, a negative regulator of c-Cbl ligase activity, turn off the ligase, and the expression level of the tyrosine kinase Syk is not affected; an enhanced formation of CIN85/Cbl complexes affects the action of Sts1 as negative regulator of Cbl turning on the E3 ligase and contributing to Syk instability.

The molecular mechanism has been elucidated by the demonstration of the involvement of the Ub-proteasome pathway in regulating Syk levels in unstimulated basophils isolated from nonreleaser donors [85]. Although no obvious differences in Cbl protein levels between releaser and nonreleaser basophils has been observed, it is possible that alterations in Cbl ligase activity also contributes to Syk instability in the nonreleaser phenotype. In this regard, we have recently demonstrated that Cbl ligase activity in unstimulated RBL-2H3 cells is influenced by relative protein abundance of specific positive and negative regulators [86] (Figure 5).

c-Cbl ligase activity is negatively regulated by the interaction with different families of scaffold proteins [87], including the Suppressor of T-cell receptor Signalling 1 (Sts1) that through its SH3 domain binds the proline-rich domain of Cbl [88]. We have demonstrated that CIN85 compete with Sts1 for Cbl binding, and that CIN85 overexpression results in a positive regulation of Cbl ligase activity promoting an antigen-independent Syk ubiquitination and degradation with consequent functional defects. It will be important to verify in the future whether the integrity of this molecular machinery is affected in non releaser basophils.

Conclusion

The complex and highly regulated apparatus leading to basophil activation is counterbalanced by an equally sophisticated series of inhibitory mechanisms.

In this chapter, we have provided a summary of the mechanisms through which Cbl proteins control the intensity and duration of FcεRI-generated signals.

Cbl-dependent receptor down-modulation occurs mainly by mean of Cbl ligase activity that promotes FcεRI β and γ multiubiquitination providing signals for receptor internalization and sorting into endocytic compartments, a process required for receptor degradation. On the other hand, Cbl interaction with adaptors involved in clathrin-mediated endocytosis allows down-modulation of engaged FcεRI through a pathway that is functionally separable from Cbl ligase activity.

Furthermore, several evidence support a role for Cbl proteins and the Ub-proteasome pathway in regulating the stability of the tyrosine kinase Syk, which plays an obligatory role in FcεRI mediated-signaling.

In conclusion, Cbl ligase and adaptor activities, by regulating the half-life of activated receptor complexes and cytoplasmic PTKs, may contribute to the down-regulation of FcεRI-mediated signaling leading to basophil effector functions.

REFERENCES

[1] Galli, S. J. (2000). Mast cells and basophils. *Curr Opin Hematol, 7,* 32-39.

[2] Prussin, C. & Metcalfe, D. D. (2003). IgE, mast cells, basophils, and eosinophils. *J Allergy Clin Immunol, 111,* 486-494.

[3] Metzger, H. (1992). The receptor with high affinity for IgE. *Immunol Rev, 125,* 37-48.

[4] Nadler, M. J., Matthews, S. A., Turner, H. & Kinet, J. P. (2000). Signal transduction by the high-affinity immunoglobulin E receptor FcεRI: coupling form to function. *Adv Immunol, 76,* 325-355.

[5] Siraganian, R. P. (2003). Mast cell signal transduction from the high-affinity IgE receptor. *Curr Opin Immunol, 15,* 639-646.

[6] Falcone, F. H., Haas, H. & Gibbs, B. F. (2000). The human basophil: a new appreciation of its role in immune responses. *Blood, 96,* 4028-4038.

[7] Kraft, S. & Kinet, J. P. (2007). New developments in FcεRI regulation, function and inhibition. *Nat Rev Immunol, 7*, 365-378.

[8] MacGlashan, D. W. Jr. (2008). IgE receptor and signal transduction in mast cells and basophils. *Curr Opin Immunol, 20,* 717-723.

[9] Molfetta, R., Peruzzi, G., Santoni, A. & Paolini, R. (2007). Negative signals from FcεRI engagement attenuate mast cell functions. *Arch Immunol Ther Exp, 55,* 219-229.

[10] Thien, C. B. & Langdon, W. Y. (2001). Cbl: many adaptations to regulate protein tyrosine kinases. *Nat Rev Mol Cell Biol, 2,* 294-307.

[11] Rao, N., Dodge, I. & Band, H. (2002). The Cbl family of ubiquitin ligases: critical negative regulators of tyrosine kinase signaling in the immune system. *J Leukoc Biol, 71,* 753-763.

[12] Dikic, I., Szymkiewicz, I. & Soubeyran, P. (2003). Cbl signaling networks in the regulation of cell function. *Cell Mol Life Sci, 60,* 1805-1827.

[13] Swaminathan, G. & Tsygankov, A. Y. (2006). The Cbl family proteins: ring leaders in regulation of cell signaling. *J Cell Physiol, 209*, 21-43.

[14] Paolini, R., Molfetta, R., Beitz, L. O., Zhang, J., Scharenberg, A. M., Piccoli, M., Frati, L., Siraganian, R. & Santoni, A. (2002). Activation of Syk tyrosine kinase is required for c-Cbl-mediated ubiquitination of FcεRI and Syk in RBL cells. *J Biol Chem, 277,* 36940-36947.

[15] Kyo, S., Sada, K., Qu, X., Maeno, K., Miah, S. M., Kawauchi-Kamata, K. & Yamamura, H. (2003). Negative regulation of Lyn protein-tyrosine kinase by c-Cbl ubiquitin-protein ligase in FcεRI-mediated mast cell activation. *Genes Cells, 8*, 825-836.

[16] Qu, X., Sada, K., Kyo, S., Maeno, K., Miah, S. M. & Yamamura, H. (2004). Negative regulation of FcεRI-mediated mast cell activation by the ubiquitin-protein ligase Cbl-b. *Blood, 103,* 1779-1786.

[17] Molfetta, R., Belleudi, F., Peruzzi, G., Morrone, S., Leone, L., Dikic, I., Piccoli, M., Frati, L., Torrisi, M. R., Santoni, A. & Paolini, R. (2005). CIN85 regulates the ligand-dependent endocytosis of the IgE receptor: a new molecular mechanism to dampen mast cell function. *J Immunol, 175,* 4208-4216.

[18] Saini, S. S., Richardson, J. J., Wofsy, C., Lavens-Phillips, S., Bochner, B. S. & Macglashan, D. W. Jr. (2001). Expression and modulation of FcεRIα and FcεRIβ in human blood basophils. *J Allergy Clin Immunol, 107,* 832-841.

[19] Donnadieu, E., Jouvin, M. H. & Kinet, J. P. (2000). A second amplifier function for the allergy-associated FcεRI-β subunit. *Immunity, 12,* 515-523.

[20] Kawakami, T. & Galli, S. J. (2002). Regulation of mast-cell and basophil function and survival by IgE. *Nat Rev Immunol, 2*, 773-786.

[21] Malveaux, F. J., Conroy, M. C., Adkinson, N. F. Jr. & Lichtenstein, L. M. (1978). IgE receptors on human basophils. Relationship to serum IgE concentration. *J Clin Invest, 62,* 176-181.

[22] Sihra, B. S., Kon, O. M., Grant, J. A. & Kay, A. B. (1997). Expression of high-affinity IgE receptors (FcεRI) on peripheral blood basophils, monocytes, and eosinophils in atopic and nonatopic subjects: relationship to total serum IgE concentrations. *J Allergy Clin Immunol, 99,* 699-706.

[23] Saini, S. S., Klion, A. D., Holland, S. M., Hamilton, R. G., Bochner, B. S. & Macglashan, D. W. Jr. (2000). The relationship between serum IgE and surface levels of FcεR on human leukocytes in various diseases: correlation of expression with FcεRI on basophils but not on monocytes or eosinophils. *J Allergy Clin Immunol, 106*, 514-520.

[24] Furuichi, K., Rivera, J. & Isersky, C. (1985). The receptor for immunoglobulin E on rat basophilic leukemia cells: effect of ligand binding on receptor expression. *Proc Natl Acad Sci U S A, 82*, 1522-1525.

[25] Lantz, C. S., Yamaguchi, M., Oettgen, H. C., Katona, I. M., Miyajima, I., Kinet, J. P. & Galli, S. J. (1997). IgE regulates mouse basophil FcεRI expression in vivo. *J Immunol, 158,* 2517-2521.

[26] MacGlashan,. D. Jr., Lichtenstein, L. M., McKenzie-White, J., Chichester, K., Henry, A. J., Sutton B. J., Gould, H. J. (1999). Upregulation of FcεRI on human basophils by IgE antibody is mediated by interaction of IgE with FcεRI. *J Allergy Clin Immunol, 104*, 492-498.

[27] MacGlashan, D. W. Jr,, Bochner, B. S., Adelman, D. C., Jardieu, P. M.,

Togias, A., McKenzie-White J., Sterbinsky, S. A., Hamilton, R. G. & Lichtenstein, L. M. (1997). Down-regulation of FcεRI expression on human basophils during in vivo treatment of atopic patients with anti-IgE antibody. *J Immunol, 158*, 1438-1445.

[28] Lin, S., Cicala, C., Scharenberg, A. M. & Kinet J. P. (1996). The FcεRIβ subunit functions as an amplifier of FcεRIγ-mediated cell activation signals. *Cell, 85,* 985-995.

[29] Simons, K. & Toomre, D. (2000). Lipid rafts and signal transduction. *Nat Rev Mol Cell Biol, 1,* 31-39.

[30] Field, K. A., Holowka, D. & Baird, B. (1997). Compartmentalized activation of the high affinity immunoglobulin E receptor within membrane domains. *J Biol Chem, 272,* 4276-4280.

[31] Sheets, E. D., Holowka, D. & Baird, B. (1999). Critical role for cholesterol in Lyn-mediated tyrosine phosphorylation of FcRI and their association with detergent-resistant membranes. *J Cell Biol, 145,* 877-887.

[32] Young, R. M., Holowka, D. & Baird, B. (2003). A lipid raft environment enhances Lyn kinase activity by protecting the active site tyrosine from dephosphorylation. *J Biol Chem, 278,* 20746-20752.

[33] Oliver, J. M., Burg, D. L., Wilson, B. S., McLaughlin, J. L. & Geahlen, R. L. (1994). Inhibition of mast cell FcεR1-mediated signalling and effector function by the Syk-selective inhibitor, piceatannol. *J Biol Chem, 269,* 29697-29703.

[34] Costello, P. S., Turner, M., Walters, A. E., Cunningham, C. N., Bauer, P. H., Downward, J. & Tybulewicz, V. L. (1996). Critical role for the tyrosine kinase Syk in signaling through the high affinity IgE receptor of mast cells. *Oncogene, 13,* 2595-2605.

[35] Zhang, J., Berenstein, E. H., Evans, R. L. & Siraganian, R. P. (1996). Transfection of Syk protein tyrosine kinase reconstitutes high affinity IgE receptor-mediated degranulation in a Syk-negative variant of rat basophilic leukemia RBL-2H3 cells. *J Exp Med, 184,* 71-79.

[36] Moriya, K., Rivera, J., Odom, S., Sakuma, Y., Muramato, K., Yoshiuchi, T., Miyamoto, M. & Yamada, K. (1997). ER-27319, an acridone-related compound, inhibits release of antigen-induced allergic mediators from mast cells by selective inhibition of Fcepsilon receptor I-mediated activation of Syk. *Proc Nat Acad Sci U S A, 94,* 12539-12544.

[37] Kepley, C. L., Youssef, L., Andrews, R. P., Wilson, B. S. & Oliver, J. M. (1999). Syk deficiency in nonreleaser basophils. *J Allergy Clin Immunol, 104,* 279-284.

[38] Lavens-Phillips, S. E. & MacGlashan, D. W. Jr. (2000). The tyrosine

kinases p53/56lyn and p72syk are differentially expressed at the protein level but not at the messenger RNA level in nonreleasing human basophils. *Am J Respir Cell Mol Biol, 266,* 566-571.

[39] MacGlashan, D. W. Jr. (2007). Relationship between spleen tyrosine kinase and phosphatidylinositol 5' phosphatase expression and secretion from human basophils in the general population. *J Allergy Clin Immunol, 119,* 626-633.

[40] Parravicini, V., Gadina, M., Kovarova, M., Odom, S., Gonzalez-Espinosa, C., Furumoto, Y., Saitoh, S., Samelson, L. E., O'Shea, J. J. & Rivera, J. (2002). Fyn kinase initiates complementary signals required for IgE-dependent mast cell degranulation. *Nat Immunol, 3,* 741-748.

[41] Gibbs, B. F. & Grabbe, J. (1999). Inhibitors of PI 3-kinase and MEK kinase differentially affect mediator secretion from immunologically activated human basophils. *J Leukoc Biol, 65,* 883-890.

[42] Miura, K. & MacGlashan, D. W. (2000). Phosphatidylinositol-3 kinase regulates p21ras activation during IgE-mediated stimulation of human basophils. *Blood, 96,* 2199-2205.

[43] Katz, H. R. (2002). Inhibitory receptors and allergy. *Curr Opin Immunol, 14,* 698-704.

[44] Bruhns, P., Frémont, S. & Daëron, M. (2005). Regulation of allergy by Fc receptors. *Curr Opin Immunol, 17,* 662-669.

[45] Kepley, C. L., Cambier, J. C., Morel, P. A., Lujan, D., Ortega, E., Wilson, B. S. & Oliver, J. M. (2000). Negative regulation of FcεRI signaling by FcγRII costimulation in human blood basophils. *J Allergy Clin Immunol, 106,* 337-348.

[46] Kimura, T., Zhang, J., Sagawa, K., Sakaguchi, K., Appella, E. & Siraganian, R. P. (1997). Syk-independent tyrosine phosphorylation and association of the protein tyrosine phosphatases SHP-1 and SHP-2 with the high affinity IgE receptor. *J Immunol, 159,* 4426-4434.

[47] Kimura, T., Sakamoto, H., Appella, E. & Siraganian, R. P. (1997). The negative signaling molecule SH2 domain-containing inositol-polyphosphate 5-phosphatase (SHIP) binds to the tyrosine-phosphorylated beta subunit of the high affinity IgE receptor. *J Biol Chem, 272,* 13991-13996.

[48] Vonakis, B. M., Gibbons, S. Jr., Sora, R., Langdon, J. M. & MacDonald, S. M. (2001). Src homology 2 domain-containing inositol 5' phosphatase is negatively associated with histamine release to human recombinant histamine-releasing factor in human basophils. *J Allergy Clin Immunol, 108,* 822-831.

[49] Thien, C. B. & Langdon, W. Y. (2005). c-Cbl and Cbl-b ubiquitin ligases:

substrate diversity and the negative regulation of signalling responses. *Biochem J, 391*, 153-166.

[50] Meng, W., Sawasdikosol, S., Burakoff, S. J. & Eck, M. J. (1999). Structure of the amino-terminal domain of Cbl complexed to its binding site on ZAP-70 kinase. *Nature, 398*, 84-90.

[51] Zheng, N., Wang, P., Jeffrey, P. D. & Pavletich, N. P. (2000). Structure of a c-Cbl-UbcH7 complex: RING domain function in ubiquitin-protein ligases. *Cell, 102*, 533-539.

[52] Joazeiro, C. A. & Weissman, A. M. (2000). RING finger proteins: mediators of ubiquitin ligase activity. *Cell, 102,* 549-552.

[53] Ciechanover, A. (1998). The ubiquitin-proteasome pathway: on protein death and cell life. *EMBO J, 17*, 7151-7160.

[54] Laney, J. D. & Hochstrasser, M. (1999). Substrate targeting in the ubiquitin system. *Cell, 97,* 427-430.

[55] Weissman, A. M. (2001). Themes and variations on ubiquitylation. *Nat Rev Mol Cell Biol, 2,* 169-178.

[56] Kassenbrock, C. K. & Anderson, S. M. (2004). Regulation of ubiquitin protein ligase activity in c-Cbl by phosphorylation-induced conformational change and constitutive activation by tyrosine to glutamate point mutations. *J Biol Chem, 279*, 28017-28027.

[57] Thrower, J. S., Hoffman, L., Rechesteiner, M. & Pickart, C. M. (2000). Recognition of the polyubiquitin proteolytic signal. *EMBO J, 19,* 94-102.

[58] Hicke, L. & Dunn, R. (2003). Regulation of membrane protein transport by ubiquitin and ubiquitin-binding proteins. *Annu Rev Cell Dev Biol, 19*, 141-172.

[59] Haglund, K., Sigismund, S., Polo, S., Szymkiewicz, I., Di Fiore, P. P. & Dikic, I. (2003). Multiple monoubiquitination of RTKs is sufficient for their endocytosis and degradation. *Nat Cell Biol, 5,* 461-466.

[60] Mosesson, Y., Shtiegman, K., Katz, M., Zwang, Y., Vereb, G., Szollosi, J. & Yarden, Y. (2003). Endocytosis of receptor tyrosine kinases is driven by mono-, not poly-ubiquitylation. *J Biol Sci, 278,* 31323-31326.

[61] Longva, K. E., Blystad, F. D., Stang, E., Larsen, A. M., Johannessen, L. E. & Madshus, I. H. (2002). Ubiquitination and proteasomal activity is required for transport of the EGF receptor to inner membranes of multivesicular bodies. *J Cell Biol, 156,* 843-854.

[62] Duan, L., Miura, Y., Dimri, M., Majumder, B., Dodge, I. L., Reddi, A. L., Ghosh, A., Fernandes, N., Zhou, P., Mullane-Robinson, K., Rao, N., Donoghue, S., Rogers, R. A., Bowtell, D., Naramura, M., Gu, H., Band, V. & Band, H. (2003). Cbl-mediated ubiquitinylation is required for lysosomal

sorting of epidermal growth factor receptor but is dispensable for endocytosis. *J Biol Chem, 278*, 28950-28960.

[63] Stang, E., Blystad, F. D., Kazazic, M., Bertelsen, V., Brodahl, T., Raiborg, C., Stenmark, H. & Madshus, I. H. (2004). Cbl-dependent ubiquitination is required for progression of EGF receptors into clathrin-coated pits. *Mol Biol Cell, 15*, 3591-3604.

[64] Rao, N., Ghosh, A. K., Ota, S., Zhou, P., Reddi, A. L., Hakezi, K., Druker, B. K., Wu, J. & Band, H. (2001). The non-receptor tyrosine kinase Syk is a target of Cbl-mediated ubiquitylation upon B-cell receptor stimulation. *EMBO J, 20*, 7085-7095.

[65] Wang, H. Y., Altman, Y., Fang, D., Dai, Y., Shao, Y. & Liu, Y. C. (2001). Cbl promotes ubiquitination of the T cell receptor ζ through an adaptor function of Zap-70. *J Biol Chem, 276,* 26004-26011.

[66] Ota, Y., Beitz, L. O., Scharenberg, A. M., Donovan, J. A., Kinet J. P. & Samelson, L. E. (1996). Characterization of Cbl tyrosine phosphorylation and a Cbl-Syk complex in RBL-2H3 cells. *J Exp Med, 184,* 1713-1723.

[67] MacGlashan, D. & Miura, K. (2004). Loss of syk kinase during IgE-mediated stimulation of human basophils. *J Allergy Clin Immunol, 114,* 1317-1324.

[68] Ota, Y. & Samelson, L. E. (1997). The product of the proto-oncogene c-cbl: a negative regulator of the Syk tyrosine kinase. *Science, 276,* 418-420.

[69] Paolini, R. & Kinet, J. P. (1993). Cell surface control of the multiubiquitination and deubiquitination of high-affinity immunoglobulin E receptors. *EMBO J, 12,* 779-786.

[70] Lafont, F. & Simons, K. (2001). Raft-partitioning of the ubiquitin ligases Cbl and Nedd4 upon IgE-triggered cell signaling. *Proc Natl Acad Sci U S A, 98,* 3180-3184.

[71] Molfetta, R., Gasparrini, F., Peruzzi, G., Vian, L., Piccoli, M., Frati, L., Santoni, A. & Paolini, R. (2009). Lipid raft-dependent FcεRI ubiquitination regulates receptor endocytosis through the action of Ubiquitin Binding Adaptors. *Submitted PLOS One.*

[72] Lu, Q., Hope, L. W., Brasch, M., Reinhard, C. & Cohen, S. N. (2003). TSG101 interaction with HRS mediates endosomal trafficking and receptor down-regulation. *Proc Natl Acad Sci USA, 100,* 7626-7631.

[73] Bache, K. G., Brech, A., Mehlum, A. & Stenmark, H. (2003). Hrs regulates multivesicular body formation via ESCRT recruitment to endosomes. *J Cell Biol, 162*, 435-442.

[74] Fattakhova, G., Masilamani, M., Borrego, F., Gilfillan, A. M., Metcalfe, D. D. & Coligan, J. E. (2006). The high-affinity immunoglobulin-E receptor

(FcRI) is endocytosed by an AP-2/clathrin-independent, dynamin-dependent mechanism. *Traffic, 7,* 673-685.

[75] Soubeyran, P., Kowanetz, K., Szymkiewicz, I., Langdon, W. Y. & Dikic, I. (2002). Cbl-CIN85-endophilin complex mediates ligand-induced downregulation of EGF receptors. *Nature, 416,* 183-187.

[76] Petrelli, A., Gilestro, G. F., Lanzardo, S., Comoglio, P. M., Migone, N. & Giordano, S. (2002). The endophilin-CIN85-Cbl complex mediates ligand-dependent downregulation of c-Met. *Nature, 416,* 187-190.

[77] Dikic, I. (2002). CIN85/CMS family of adaptor molecules. *FEBS Lett, 529,* 110-115.

[78] Lupher, M. L. Jr., Rao, N., Lill, N. L., Andoniou, C. E., Miyake, S., Clark, E. A., Druker, B. & Band, H. (1998). Cbl-mediated negative regulation of the Syk tyrosine kinase. A critical role for Cbl phosphotyrosine-binding domain binding to Syk phosphotyrosine 323. *J Biol Chem, 273,* 35273-35281.

[79] Ota, S., Hazeki, K., Rao, N., Lupher, M. L. Jr., Andoniou, C. E., Druker, B. & Band, H. (2000). The RING finger domain of Cbl is essential for negative regulation of the Syk tyrosine kinase. *J Biol Chem, 275,* 414-422.

[80] Andoniou, C. E., Lill, N. L., Thien, C. B., Lupher, M. L. Jr., Ota, S., Bowtell, D. D., Scafe, R. M., Langdon, W. Y. & Band, H. (2000). The Cbl proto-oncogene product negatively regulates the Src-family tyrosine kinase Fyn by enhancing its degradation. *Mol Cell Biol, 20*, 851-867.

[81] Rao, N., Miyake, S., Reddi, A. L., Douillard, P., Ghosh, A. K., Dodge, I. L., Zhou, P., Fernandes, N. D. & Band, H. (2002). Negative regulation of Lck by Cbl ubiquitin ligase. *Proc Natl Acad Sci U S A, 99,* 3794-3799.

[82] Sohn, H. W., Gu, H. & Pierce, S. K. (2003). Cbl-b negatively regulates B cell antigen receptor signaling in mature B cells through ubiquitination of the tyrosine kinase Syk. *J Exp Med, 197,* 1511-1524.

[83] MacGlashan, D. W. Jr., Ishmael, S., MacDonald, S. M., Langdon, J. M., Arm, J. P. & Slogane, D. E. (2008). Induced loss of Syk in human basophils by non-IgE-dependent stimuli. *J Immunol, 180*, 4208-4217.

[84] Kepley, C. L., Youssef, L., Andrews, R. P., Wilson, B. S. & Oliver, J. M. (2000). Multiple defects in FcεRI signaling in Syk-deficient nonreleaser basophils and IL-3-induced recovery of Syk expression and secretion. *J Immunol, 165*, 5913-5920.

[85] Youssef, L. A., Wilson, B. S. & Oliver, J. M. (2002). Proteasome-dependent regulation of Syk tyrosine kinase levels in human basophils. *J Allergy Clin Immunol, 110,* 366-373.

[86] Peruzzi, G., Molfetta, R., Gasparrini, F., Vian, L., Morrone, S., Piccoli, M.,

Frati, L., Santoni, A. & Paolini, R. (2007). The adaptor molecule CIN85 regulates Syk tyrosine kinase level by activating the ubiquitin-proteasome degradation pathway. *J Immunol, 179,* 2089-2096.

[87] Ryan, P. E., Davies, G. C., Nau, M. M. & Lipkowitz, S. (2006). Regulating the regulator: negative regulation of Cbl ubiquitin ligases. *Trends Biochem Sci, 31*, 79-88.

[88] Kowanetz, K., Corsetto, N., Haglund, K., Schmidt, M. H., Heldin, C. H. & Dikic, I. (2004). Suppressor of T-cell receptor signalling Sts-1 and Sts-2 bind to Cbl and inhibit endocytosis of receptor tyrosine kinase. *J Biol Chem, 279,* 32786-32795.

In: Basophil Granulocytes
Editor: Paul K. Vellis
ISBN: 978-1-60741-797-2

Chapter 2

BASOPHILS, MAST CELLS AND LIVER CANCER

*Fabio Grizzi**
Istituto Clinico Humanitas IRCCS, Rozzano, Milan, Italy.

ABSTRACT

Liver cancer is still a major public health problem worldwide. It is now known that solid tumors, including hepatocellular carcinoma, are infiltrated by immune cells most prominently T and B lymphocytes. All of them are variably scattered within the tumor and loaded with an assorted array of cytokines, chemokines, and inflammatory and cytotoxic mediators. This complex network reflects the diversity in tumor biology and tumor-host interactions.

Basophil granulocytes and mast cells are well established effector cells in IgE-associated immune responses. While in allergies or parasitic infections the role of these cells has been known for years, in cancer there are what seem to be conflicting data describing a supporting or a possibly inhibitory role of these cells in several human neoplasia.

Here, these knowledge will summarize, focusing on the implications of these findings in the understanding the role of basophil granulocytes and mast cells in primary and secondary liver cancer.

* Corresponding author: Email: fabio.grizzi@humanitas.it

INTRODUCTION

Human carcinogenesis is a dynamical process that depends on a high number of variables and is regulated at multiple spatial and temporal scales [1,2]. Despite advances in our genomic and cellular knowledge, hepatocellular carcinoma (HCC) remains one of the major public health problems throughout the world [3-7]. HCC represents the fifth common cancer worldwide [5]. It is now recognized to be highly heterogeneous: it encompasses a wide range of clinical behaviors, and is underpinned by a complex array of gene alterations that affect molecular, cellular and supra-cellular processes [8].

Solid tumors, including HCC, are commonly infiltrated by a high number of immune cells [9]. All of them are variably scattered within the tumor and loaded with an assorted array of cytokines, chemokines, and inflammatory and cytotoxic mediators. This complex network reflects the diversity in tumor biology and tumor-host interactions.

Basophil granulocytes and mast cells are well established effector cells in IgE-associated immune responses [10]. Although basophil granulocytes share some properties with tissue mast cells, such as basophilic staining, histamine content and the expression of high levels of the high-affinity IgE receptor, they are actually recognized as specialized blood cells most closely related to eosinophils [10-17]. While in allergies or parasitic infections the role of basophils and mast cells has been known for years, in cancer there are what seem to be conflicting data describing a supporting or a possibly inhibitory role of these cells in several human neoplasia.

Here we review these knowledge, focusing on the implications of these findings in the understanding the role of basophil granulocytes and mast cells in primary and secondary liver cancer.

Liver Cancer: A Still Complex Disease

HCC is the most common primary cancer associated with hepatitis B (HBV) and hepatitis C virus (HCV) chronic infections, alcohol abuse or in developing countries with food contaminated with *Aspergillus flavus* fungus [5-7,18]. The incidence of HCC has been rising in the last two decades in Europe, United states and Japan [5,7,19]. The number of new cases is estimated to be 564,000 per year. About 80% of all cases are found in Asia [20].

Despite advances in our cellular and molecular knowledge HCC remains one of the major public health problems throughout the world. It is now known to be highly heterogeneous [8, 21-24]. A variety of genomic and molecular alterations have been identified in fully developed HCC and to a lesser extent in morphologically defined precancerous lesions [21-24]. However, this database has been obtained largely in a shatter and uncoordinated manner, typically in investigations in which aberrations in single genes have been analyzed in a group of HCC from humans and experimental models. Although these data all seems to be helpful, they have not led to a coherent understanding of the complex mechanisms underlying HCC development, or to the identification of critical genomic and/or molecular aberrations that improve the carefulness of diagnosis or therapeutic interventions.

HCC is a *non-linear* complex disease that emerges from multiple spontaneous and/or inherited mutations that induce dramatic changes in expression patterns of genes and proteins that function in networks controlling critical cellular events [1].

Therefore, it is compulsory to identify HCC and the recurrence at its earlier period. A number of serum markers have been proposed and currently employed as an effective method for detecting HCC for a long time [25]. Four categories of HCC tumor markers have mainly been proposed and currently applied in the clinical practice. In addition, a family of tumor-associated antigens called cancer-testis (CT) antigens has recently been identified and their encoding genes have been extensively investigated. CT antigens are being vigorously pursued as targets for therapeutic cancer vaccines [26,27].

Mast Cells: A Heterogeneous Cell Population

Mast cells have a rather unique position among cells of the immune response. Mast cells were first described by Paul Ehrlich in his 1878 doctoral thesis: he called them "mastzellen" (*maestung* - a root of the English word mastication; the active form "measten" is still in use) because of their characteristic staining of proteoglycan and protease-rich cytoplasmic granules [29]. Ehrlich also noted the tendency of mast cells to be associated with blood vessels, nerves, and glandular ducts.

It has been estimated that human mast cells contain 2.4 to 7.8 μg heparin per 10^6 cells [30]. This observation, along with the knowledge that heparin is a negatively charged molecule helps explain why mast cells granules are preferentially stained with cationic dyes [31].

It has long been recognized that mast cells elicit allergic symptoms [32], but it is now widely accepted that they are multifunctional effector cells of the immune system, although the various phases of their differentiation are still only partially known. It was initially suggested that they derive from T lymphocytes, fibroblasts or macrophages, but the current general consensus suggests that they originate from pluripotent hemopoietic stem cells in bone marrow, from where they are released into the blood as progenitors before they undergo terminal differentiation by invading connective or mucosal tissue as morphologically unidentifiable mast cells precursors and then differentiating into phenotypically identifiable mast cells [31]. It has been demonstrated that mast cells are long-lived cells [31].

Mast cells are found in almost all of the major organs of the human body, and in a large number of body sites that come into contact with the external environment, including the skin, respiratory system and digestive tract. These main accumulations in sites where foreign material attempts host invasion suggest that mast cells are one of the first cell populations to initiate defense mechanisms.

The ability of the lineage to generate individual mast cells populations with different biochemical and functional properties gives them greater diversity and flexibility in meeting the requirements of the physiological, immunological, inflammatory or other biological responses in which they may be involved [32-36].

Several Authors have identified a local or systemic increase in the number of mast cells in various pathological conditions, including interstitial pneumonia, ulcerative colitis, intestinal helminthosis and ectodermal parasitosis, as well as skin disorders such as atopic dermatitis, psoriasis, scleroderma and wound healing, and various neoplastic diseases [35-41].

Tryptases and chymases are the major protein components of mast cells secretory granules. TNF was the first cytokine clearly associated with normal mast cells in 1990 [42]. Other mast cells products include interleukins (IL-3, IL-4, IL-5, IL-6, IL-9, IL-10, IL-11, IL-12, IL-13, IL-15, IL-16, IL-18, IL-2), chemokines (macrophage inflammatory protein alpha [MIP-1]), hematopoietic factors (granulocyte macrophage colony stimulating factor [GM-CSF]), stem cell factor (SCF), growth factors (transforming growth factor beta [TGF-β], vascular endothelial growth factor [VEGF], nerve growth factor [NGF]), several metalloproteinases, heparin, histamine, chondroitin sulfates, cathepsin, carboxypeptidases and peroxidase [35,43].

These products may be released when mast cells are activated via IgE- or IgG-dependent mechanisms, and may also be produced under other circumstances such as in response to stimulation by bacterial products through Toll-like receptors (TLRs) [44].

Mast cells and IgE have long been associated with the pathogenesis of the acute manifestations of the immediate hypersensitivity reaction, the pathophysiologic hallmark of allergic rhinitis, allergic asthma, and anaphylaxis. The central role of mast cells in these disorders is now widely accepted. Additionally, mast cells are considered to be critical effectors in many human inflammatory diseases, and the core of an immediate hypersensitive reaction: they have been incriminated in different diseases including allergy, asthma, rheumatoid arthritis, arteriosclerosis, chronic graft-versus-host disease, fibrotic disease, ischemic heart disease and malignancy, and contribute to the progression of chronic diseases [45-49]. Any alteration in cell programs that determines a requirement for mast cells degranulation may therefore have a considerable impact on disease severity. Nguyen *et al.* have shown the ability of Prostaglandin E2 (PGE_2) to initiate mast cells degranulation changes in the aging animal, thus suggesting that aging induces reprogramming of mast cells degranulation [50].

Human Basophils Granulocytes: A Rare But Emerging Inflammatory Cell Population

Although peripheral blood basophil numbers are modestly elevated (approximately 2-fold) in allergic asthma, these cells make up less than 1% of circulating blood leukocytes and are present in all vertebrates [51]. Recognized originally by Paul Ehrlich by their cytoplasmic granules that stain with basophilic dyes, basophil granulocytes are typically grouped with mast cells based on their appearance in tissues during allergic and antihelminth immune responses and on their expression of high-affinity IgE receptor (Fc3RI) that renders these cell types responsive to activation by cross-linking IgE bound to the surface (Figure 1).

It is now well known that basophils develop from $CD34^+$ pluripotent stem cells, differentiate and mature in the bone marrow, and circulate in the periphery [11]. The lifespan of basophils (several days) is much shorter than that of mast cells (weeks or months), and basophil granulocytes do not proliferate after they mature [51]. Basophil granulocytes comprise a separate lineage from mast cells, although both cells share common features, including T_H2 cytokine expression, and histamine release. Morphologically, basophil granulocytes exhibit a segmented nucleus. They are usually identified by means of metachromatic staining with basic dyes, such as toluidine blue [11].

Phenotypically, basophils differ from mast cells and eosinophils in their lack of expression of the cell surface antigen c-Kit (CD117) and the chemokine receptor CCR3, respectively [12].

Basophils express a high variety of cytokine [(IL-3R, IL-5R, and granulocyte-macrophage colony-stimulating factor (GM-CSFR)], chemokine (CCR2 and CCR3), complement (CD11b, CD11c, CD35, and CD88), prostaglandin ($CRTH_2$), and immunoglobulin Fc (FcεRI and FcγRII) receptors [11,52].

IL-3 is the dominant cytokine driving basophil differentiation and is sufficient to differentiate stem cells into basophil granulocytes [11]. Basophils differentiate from a common basophil-eosinophil precursor; this is supported by the derivation of mixed colonies of basophils and eosinophils from individual precursor cells [53].

Two basophil granule-specific monoclonal antibodies, called BB-1 and 2D7, have recently been developed, permitting the identification of basophil granulocytes in tissues and greatly furthering our understanding of the role of basophils in allergic diseases and asthma [11, 54].

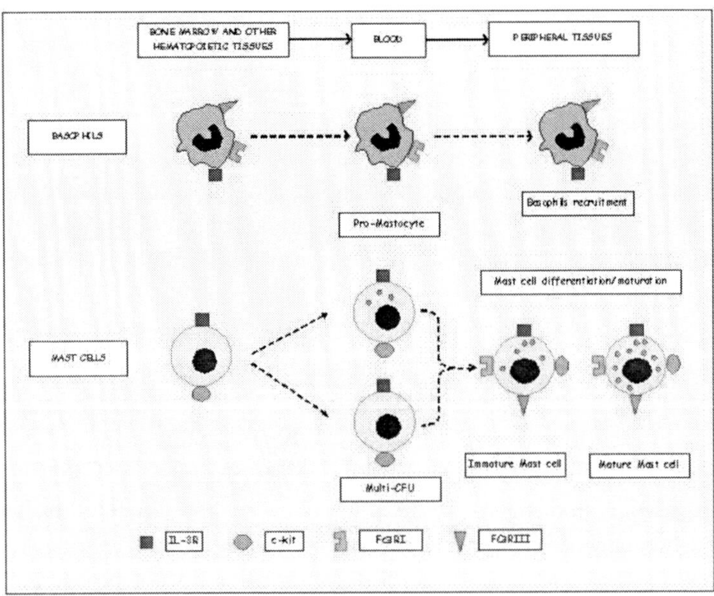

Figure 1. A simplified schema showing the dynamics of development and maturation of basophils granulocytes and mast cells. Basophils originate from multipotential hematopoietic progenitor cells, but mature in the bone marrow before entering the peripheral circulation. Mast cells originate from multipotential hematopoietic progenitor cells, but complete major parts of their differentiation and maturation process in the colonizing tissues.

Basophils express a complete and functional FcεRI receptor (αβγ2), cross-linking of which leads to basophil activation, granule exocytosis, and mediator release [55]. C3a and C5a also activate basophils through specific receptors. Activation leads to histamine release, eicosanoid synthesis, and IL-4/IL-13 gene expression [11]. It has been demonstrated that additional mediators, such as CC chemokines (eotaxins [CCL11, CCL24, and CCL26], monocyte chemoattractant protein 3 [CCL7], monocyte chemoattractant protein 4 [CCL13], and RANTES [CCL5]), formylmethionine-leucine-phenylalanine, IL-3, IL-5, GM-CSF, and histamine-releasing factor, do not directly cause basophil mediator release-production but potentiate FcεRI effects [11, 56]

Basophil granulocytes are a dominant and rapid source of IL-4 in both allergen- and helminth-specific responses in both humans and mice [57,58]. It is well known that IL-4 is fundamental for the differentiation of naive $CD4^+$ T cells into IL-4-producing T_H2 cells, but the cellular source of the initial IL-4 remains unknown. Basophils generate large quantities of T_H2-type cytokines in response to various stimuli, including IL-3 and parasite antigens. These results all indicated that basophils might be involved in T_H2-cell differentiation. Conversely, the mast cell mediators PGD2 and IL-5 are not produced by basophils [11].

Although uninvestigated for several years because the inability to follow these cells by other than morphologic criteria, it seems now clear that basophils may provide unique functions unmet by other hematopoietic cells, particularly during T_H2-associated allergic and anti-helminth responses.

The physiologic role of basophil granulocytes is still not well unexplored, although presumably they fulfill a host defense function [11,59]. A role for basophils in innate immunity is suggested by their expression of a functional TLR-2 receptor [60], as well as their non–IgE-dependent activation by proteases from Der p 1 and hookworm [61]. Additionally, basophils are known to play a role in rejection of ticks and are a component of the inflammatory response to many parasites. Basophil granulocytes have been identified in cutaneous and pulmonary late-phase allergic responses and are found in increased numbers in the lungs of patients who die of asthma [11].

In a recent report, basophils are shown to increase humoral memory immune responses by producing IL-4 and IL-6 when stimulated by reexposure to the allergen to which specific IgE had been produced in the primary immune response [51].

Mast Cells, Basophil Granulocytes and the Liver Cancer

Mast cells were recognized to infiltrate the interface between developing tumors and healthy issues as early as 1891 by Westphal using early metachromatic staining techniques on primary tumors [62-64]. He recognized that mast cells tended to aggregate in the tissue adjacent to cancers rather than in the tumor itself.

Potential mast cell effects on tumor growth can be categorized as *direct effects* on tumor cells, such as mast cell-mediated cytotoxicity, or *indirect effects* such as mast cell-directed angiogenesis, tissue remodeling of the neighboring environment and immune cell recruitment.

Data on mast cell function in developing tumors has largely resulted from mouse models of cancer, with complementary, correlative studies in human patients. As such, it is important to note that important differences exist between human and mouse mast cell subsets. In mice, Mast cells are broadly distinguished as the short-lived mucosal or long-lived connective tissue subtypes based on: 1) their location 2) their complement of proteases, and 3) their growth factor requirements [63,65].

Mast cells are now recognized as an early and persistent infiltrating cell type in many tumors, often entering before significant tumor growth and angiogenesis have occurred. They have been shown to accumulate in and around adenomatous polyps (precursors to invasive colon cancer) and skin dysplasias [66] prior to tumor development and around many aggressive human tumors [67], particularly malignant melanoma [68], breast carcinoma [69] and colorectal carcinoma [70]. Much of the knowledge on mast cell function and the neoplastic disease results from findings in wound healing, allergic asthma and parasite infection.

It has been demonstrated that the infiltration of mast cells at the tumor site can greatly enhance tumorigenesis [67], although these findings are still debated. It is observed that tumor-infiltrating mast cells express multiple pro-inflammatory factors and increase IL-17 expression in tumor. Additionally, it is believed that mast cells may impact upon the growth of tumors by multiple mechanisms, including angiogenesis. Several studies have demonstrated that early angiogenic activity is dependent upon mast cells and is an essential part of neoplastic development, with MCs mediating this activity by releasing heparin, VEGF and IL-8 [67]. It is now recognized that tumors dynamically create immune tolerance against themselves by creating sites of immune privilege and by inducing anti-tumor T cell tolerance [71]. The induction of immunological ignorance, induction of *anergy* in effector T cells and active suppression of effector cells have all been reported to contribute to the lack of immune responses to tumors [67,72]. A fundamental question remains whether and how mast cells contribute to the development of immune privilege within the tumor microenvironment [67]. It is now indubitable that a major point linking mast cells to cancer is the capacity of these cells to synthesize and release potent angiogenic cytokines, such as VEGF, FGF-2, the serine proteases tryptase and chymase, IL-8, TGF-b, TNF-a and nerve growth factor (NGF) [73]. It has also demonstrated that not only mast cells stimulate tumor angiogenesis but they also promote lymphangiogenesis [73].

It has been associated mast cells and angiogenesis in liver neoplasia because the significant correlation between mast cells and microvessel density suggests they may play a role in tumor progression by promoting angiogenesis [40]. We showed that HCC contained a higher density of toluidine blue stained mast cells, thus suggesting that the recruitment of these cells increases during the development of HCC [40]. However, these preliminary results obtained from 22 HCC patients did not indicate any significant correlation between mast cells density and disease stage. We hypothesize that the lower density observed in the most severe stage may be due to the massive degranulation of a large number of mast cells.

Moreover, the obtained results allow the following conclusions to be drawn: *a)* mast cells density did not significantly correlate with disease grade, but there was a trend towards a decrease in mast cells number as the grade increased. *b)* There was no correlation between mast cells density and some baseline characteristics of the patients, such as the sex or age of the patients, since the tumor evolves more slowly in older patients, we expected to find a higher density in younger cases. This hypothesis was partially substantiated by an albeit statistically non-significant negative correlation between MCs density and age. This behavior was in line with theoretical models showing that a large number of biological events were age-dependent [75]. *c)* mast cells density did not correlate with serum ALT or AST levels. HCV, cirrhosis and HCC were associated with persistent or fluctuating elevations in ALT levels, but did not distinguish among these conditions. *d)* There was no significant difference in the density of mast cells between patients with and without HCV disease. As it is known that there is a close relationship between HCC and HCV infection, it is not surprising that we had more HCV-positive than HCV-negative cases. Nevertheless, mast cells density was similar in both groups, suggesting that HCV infection did not increase mast cell recruitment. *e)* The close proximity of mast cells and surrounding tumor cells suggested the existence of roles of mast cells in the development of HCC, including tumor growth as well as host immunity and stromal reaction. *f)* The density of mast cells was higher in the specimens showing a greater capillarisation of sinusoidal endothelial cells.

Recently Lampiasi *et al.* found that mediators stored in mast cell granules and histamine may affect the growth of liver cancer cells [76]. In addition, they suggest that histamine and histamine receptor agonists/antagonists might be considered as "new therapeutic" drugs to inhibit liver tumor growth [76].

Peng *et al.* have shown that tumor-associated macrophages and mast cells might be closely related to the enhancement of tumor angiogenesis and that mast cell density might be associated with tumor differentiation and prognosis of HCC [77].

Compared with the extensive research on the roles of mast cells, basophil granulocytes have not been thoroughly studied. This can be attributed to the paucity of basophil sources and the lack of appropriate analytical tools. Researchers carrying out biochemical and functional analyses of basophils are often faced with the difficulty of collecting them in sufficient numbers, particularly from mice. In addition, there are currently no available cell lines that are relevant to basophils. since the discovery of basophils by Paul Ehrlich at the end of the nineteenth century, the functional significance of this small population of cells in the blood has remained an enigma, even though basophils are found in most vertebrates.

CONCLUSIONS

On the basis of the above considerations the following conclusions are made:

1. It is now known that solid tumors, including HCC, are commonly infiltrated by immune cells. All of them are variably scattered within the tumor and loaded with an assorted array of cytokines, chemokines, and inflammatory and cytotoxic mediators. This complex network reflects the diversity in tumor biology and tumor-host interactions.
2. Basophil granulocytes and mast cells are well established effector cells in IgE-associated immune responses.
3. Mast cells are found in almost all of the major organs of the human body, and in a large number of body sites that come into contact with the external environment, including the skin, respiratory system and digestive tract. These main accumulations in sites where foreign material attempts host invasion suggest that mast cells are one of the first cell populations to initiate defense mechanisms.

4. Tryptases and chymases are the major protein components of mast cells secretory granules. TNF was the first cytokine clearly associated with normal mast cells in 1990 [42]. Other include interleukins, chemokines, hematopoietic factors, SCF, growth factors including TGF-β, VEGF, NGF, several metalloproteinases, heparin, histamine, chondroitin sulfates, cathepsin, carboxypeptidases and peroxidase [35,43].
5. Basophils share several striking similarities with mast cells, including expression of the high-affinity receptor for immunoglobulin E (IgE) (Fc3RI), the ability to secrete—upon stimulation with IgE and specific antigen - a similar, but not identical, spectrum of mediators and cytokines, and the presence of cytoplasmic granules that stain metachromatically with certain basic dyes [15].
6. Although mature mast cells are not normally present in the blood, but instead enter tissues as immature progenitors that undergo the last stages of their development in those anatomical sites where they ultimately will reside basophils typically mature in hematopoietic tissues and then, like neutrophils and eosinophils, circulate in the blood until they are eliminated or recruited into tissues.
7. The detection of basophils is difficult, particularly in mice, owing to their rarity (less than 1% of leukocytes in the peripheral blood, spleen and bone marrow are basophils) and because mouse basophils have fewer basophilic granules than human basophils [51].
8. It is now clear that basophils from humans and mice rapidly secrete larger quantities of interleukin-4 (IL-4) than T_H2 cells in response to various stimuli, including signaling through FcεRI.
9. Mast cells were recognized to infiltrate the interface between developing tumors and healthy issues as early as 1891 by Westphal using early metachromatic staining techniques on primary tumors [62-64]. The possible effects of mast cells on cancer might be simplified in three categories: a) mast cells-released cytokines, b) the influence of mast cells on tumor angiogenesis and c) the effect of mast cells on tumor via heparin release [74].
10. Several Authors have shown that HCC contained a higher density of toluidine blue stained mast cells, thus suggesting that the recruitment of these cells increases during the development of HCC.
11. Compared with the extensive research on the roles of mast cells, basophil granulocytes have not been thoroughly studied. This can be attributed to the paucity of basophil sources and the lack of appropriate analytical tools.

New findings suggest that mast cells may serve as a novel therapeutic target for cancer treatment and that inhibiting mast cell function may lead to tumor regression. It is also indubitable that recent knowledge, have markedly changed our understanding of basophils, and we now appreciate that they are key players in the immune system. Epidemiological studies have suggested inverse associations between allergic diseases and malignancies [78]. It has been shown that passive anaphylaxis or weekly injection of histamine and serotonin inhibited tumor growth in a transplant mouse model [78]. In allergy IgG antibodies may block IgE binding to allergens, whereas other IgG antibody specificities enhance this and support the anaphylactic reaction. In cancer, inhibitory antibody specificities prevent growth signals derived from over-expressed oncogenes, whereas growth-promoting specificities enhance signaling and proliferation [79].

In conclusion, we have learned that although a subpopulation of cells might be small, this does not necessarily mean that it does not have an important biological role. Accumulating evidence indicates that basophils have pivotal roles in the regulation of the immune system in spite of their small numbers. Therefore, basophils and their products might be promising therapeutic targets for several immunological disorders.

References

[1] Grizzi, F; Chiriva-Internati, M. Cancer: looking for simplicity and finding complexity. *Cancer Cell Int*, 2006, 6, 4.

[2] Grizzi, F; Russo, C; Portinaro, N; Hermonat PL; Chiriva-Internati M. Complexity and cancer. *Gastroenterology*, 2004, 126, 630-631.

[3] Trinchet, JC; Alperovitch, A; Bedossa, P; Degos, F; Hainaut, P; Beers, BV. Epidemiology, prevention, screening and diagnosis of hepatocellular carcinoma. *Bull Cancer*, 2009, 96, 35-43.

[4] Okuda, H. Hepatocellular carcinoma development in cirrhosis. *Best Pract Res Clin Gastroenterol*, 2007, 21, 161-173.

[5] Hui, KM. Human hepatocellular carcinoma: Expression profiles-based molecular interpretations and clinical applications. *Cancer Lett*, 2008. [Epub ahead of print]

[6] Zerbini, A; Pilli, M; Ferrari, C; Missale, G. Is there a role for immunotherapy in hepatocellular carcinoma? *Dig Liver Dis*, 2006, 38, 221-225.

[7] Jemal, A; Murray, T; Ward, E; Samuels, A; Tiwari, RC; Ghafoor, A; Feuer, EJ; Thun, MJ. Cancer statistics, 2005. *CA Cancer J Clin*, 2005, 55, 10-30.

[8] Thorgeirsson, SS; Lee, JS; Grisham, JW. Functional genomics of hepatocellular carcinoma. *Hepatology*, 2006, 43, S145-S150.

[9] Berasain, C; Castello, J; Perugorria, MJ; Latasa, MU; Prieto, J; Avila, MA. Inflammation and liver cancer: new molecular links. *Ann N Y Acad Sci*, 2009, 1155, 206-221.

[10] Marone, G; Lichtenstein, LM; Galli, SJ. Mast cells and Basophils. London, San Diego, Academic Press, 2000.

[11] Prussin, C; Metcalfe, DD. IgE, mast cells, basophils, and eosinophils. *J Allergy Clin Immunol*, 2006, 117, S450-S456.

[12] Min, B. Basophils: what they 'can do' versus what they 'actually do'. *Nat Immunol*, 2008, 9, 1333-1339.

[13] Min, B; Le Gros, G; Paul, WE. Basophils: a potential liaison between innate and adaptive immunity. *Allergol Int*, 2006, 55, 99-104.

[14] Sullivan, BM; Locksley, RM. Basophils: a nonredundant contributor to host immunity. *Immunity*, 2009, 30, 12-20.

[15] Galli, SJ; Franco, CB. Basophils are back! *Immunity*, 2008, 28, 495-497.

[16] Kawakami, T; Galli, SJ. Regulation of mast-cell and basophil function and survival by IgE. *Nat Rev Immunol*, 2002, 2, 773-786.

[17] Galli, SJ. Mast cells and basophils. *Curr Opin Hematol*, 2000, 7, 32-39.

[18] Jaskiewicz, K; Stepien, A; Banach, L. Hepatocellular carcinoma in a rural population at risk. *Anticancer Res*, 1991, 11, 2187-2189.

[19] Tanaka, Y; Hanada, K; Mizokami, M; Yeo, AE; Shih, JW; Gojobori, T; Alter, HJ. A comparison of the molecular clock of hepatitis C virus in the United States and Japan predicts that hepatocellular carcinoma incidence in the United States will increase over the next two decades. *Proc Natl Acad Sci U S A*, 2002, 99, 15584-15589.

[20] Lai, EC; Lau, WY. The continuing challenge of hepatic cancer in Asia. *Surgeon*, 2005, 3, 210-215.

[21] Llovet, JM; Burroughs, A; Bruix, J. Hepatocellular carcinoma. *Lancet*, 2003, 362, 1907-1917.

[22] Kojiro, M. Histopathology of liver cancers. *Best Pract Res Clin Gastroenterol*, 2005, 19, 39-62.

[23] Lee, JS; Thorgeirsson, SS. Genome-scale profiling of gene expression in hepatocellular carcinoma: classification, survival prediction, and identification of therapeutic targets. *Gastroenterology*, 2004, 127, S51-S55.

[24] Thorgeirsson, SS; Grisham, JW. Molecular pathogenesis of human hepatocellular carcinoma. *Nat Genet*, 2002, 31, 339-346.

[25] Zhou, L; Liu, J; Luo, F. Serum tumor markers for detection of hepatocellular carcinoma. *World J Gastroenterol*, 2006, 12, 1175-1181.

[26] Chiriva-Internati, M; Grizzi, F; Wachtel, MS; Jenkins, M; Ferrari, R; Cobos, E; Frezza, EE. Biological treatment for liver tumor and new potential biomarkers. *Dig Dis Sci*, 2008, 53, 836-843.

[27] Grizzi, F; Franceschini, B; Hamrick, C; Frezza, EE; Cobos, E; Chiriva-Internati, M. Usefulness of cancer-testis antigens as biomarkers for the diagnosis and treatment of hepatocellular carcinoma. *J Transl Med*, 2007, 5, 3.

[28] Chiriva-Internati, M; Grizzi, F; Jumper, CA; Cobos, E; Hermonat, PL; Frezza, EE. Immunological treatment of liver tumors. *World J Gastroenterol*, 2005, 11, 6571-6576.

[29] Ehrlich, P. Beitrage zur kenntis der granulierted bindegwebszellen und der eosinophilen leukozyten. *Arch Anat Physiol*, 1879, 3, 166-169.

[30] Metcalfe, DD; Lewis, RA; Silbert, JE; Rosenberg, RD; Wasserman, SI; Austen, KF. Isolation and characterization of heparin from human lung. *J Clin Invest*, 1979, 64, 1537–1543.

[31] Metcalfe, DD. Mast cells and mastocytosis. *Blood*, 2008, 112, 946-956.

[32] Metz, M; Maurer, M. Mast cells - key effector cells in immune responses. *Trends Immunol*, 2007, 28, 234-241.

[33] Galli, SJ; Grimbaldeston, M; Tsai, M. Immunomodulatory mast cells: negative, as well as positive, regulators of immunity. *Nat Rev Immunol*, 2008, 8, 478-486.

[34] Bradding, P. Human mast cell cytokines. *Clin Exp Allergy*, 1996, 26, 13-19.

[35] Krüger-Krasagakes, S; Czarnetzki, BM. Cytokine secretion by human mast cells. *Exp Dermatol*, 1995, 4, 250-254.

[36] Gruber, BL. Mast cells in the pathogenesis of fibrosis. *Curr Rheumatol Rep*, 2003, 5, 147-153.

[37] Franceschini, B; Ceva-Grimaldi, G; Russo, C; Dioguardi, N; Grizzi, F. The complex functions of mast cells in chronic human liver diseases. *Dig Dis Sci*, 2006, 51, 2248-2256.

[38] Welle, M. Development, significance, and heterogeneity of mast cells with particular regard to the mast cell-specific proteases chymase and tryptase. *J Leukoc Biol*, 1997, 61, 233-245.

[39] Persinger, MA; Lepage, P; Simard, JP; Parker, GH. Mast cell numbers in incisional wounds in rat skin as a function of distance, time and treatment. *Br J Dermatol*, 1983, 108, 179-187.

[40] Grizzi, F; Franceschini, B; Chiriva-Internati, M; Liu, Y; Hermonat, PL; Dioguardi, N. Mast cells and human hepatocellular carcinoma. *World J Gastroenterol*, 2003, 9, 1469-1473.

[41] Racanelli, V; Rehermann, B. The liver as an immunological organ. *Hepatology*, 2006, 43, S54-S62.

[42] Gordon, JR; Galli, SJ. Mast cells as a source of both preformed and immunologically inducible TNF-alpha/cachectin. *Nature*, 1990, 346, 274-276.

[43] Gilfillan, AM; Rivera, J. The tyrosine kinase network regulating mast cell activation. *Immunol Rev*, 2009, 228, 149-69.

[44] Kulka, M; Alexopoulou, L; Flavell, RA; Metcalfe, DD. Activation of mast cells by double-stranded RNA: evidence for activation through toll-like receptor-3 (TLR-3). *J Allergy Clin Immunol*, 2004, 114, 174–182.

[45] Gagliano, N; Grizzi, F; Annoni, G. Mechanisms of aging and liver functions. *Dig Dis*, 2007, 25, 118-123.

[46] Krishnaswamy, G; Kelley, J; Johnson, D; Youngberg, G; Stone, W; Huang, SK; Bieber, J; Chi, DS. The human mast cell: functions in physiology and disease. *Front Biosci*, 2001, 6, 1109-1127.

[47] Abd-El-Aleem, SA; Morgan, C; Ferguson, MW; McCollum, CN; Ireland, GW. Spatial distribution of mast cells in chronic venous leg ulcers. *Eur J Histochem*, 2005, 49, 265-272.

[48] Yu, M; Tsai, M; Tam, SY; Jones, C; Zehnder, J; Galli, SJ. Mast cells can promote the development of multiple features of chronic asthma in mice. *J Clin Invest*, 2006, 116, 1633-1641.

[49] Sankovic, S; Dergenc, R; Bojic, P. Mast cells in chronic inflammation of the middle ear mucosa. *Rev Laryngol Otol Rhinol (Bord)*, 2005, 126, 15-18.

[50] Nguyen, M; Pace, AJ; Koller, BH. Age-induced reprogramming of mast cell degranulation. *J Immunol*, 2005, 175, 5701-5707.

[51] Karasuyama, H; Mukai, K; Tsujimura, Y; Obata, K. Newly discovered roles for basophils: a neglected minority gains new respect. *Nat Rev Immunol*, 2009, 9, 9-13.

[52] Bochner, BS; Schleimer, RP. Mast cells, basophils, and eosinophils: distinct but overlapping pathways for recruitment. *Immunol Rev*, 2001, 179, 5-15.

[53] Denburg, JA; Telizyn, S; Messner, H; Lim, B; Jamal, N; Ackerman, SJ; Gleich, GJ; Bienenstock, J. Heterogeneity of human peripheral blood eosinophil-type colonies: evidence for a common basophil-eosinophil progenitor. *Blood*, 1985, 66, 312-318.

[54] Buckley, MG; Euen, AR; Walls, AF. The return of the basophil. *Clin Exp Allergy*, 2002, 32, 8-10.

[55] Kinet, JP. The high-affinity IgE receptor (FcεRI): from physiology to pathology. *Annu Rev Immunol*, 1999, 17, 931-972.

[56] Schroeder, JT; MacGlashan, DW; Jr; Lichtenstein, LM. Human basophils: mediator release and cytokine production. *Adv Immunol*, 2001, 77, 93-122.

[57] Devouassoux, G; Foster, B; Scott, LM; Metcalfe, DD; Prussin, C. Frequency and characterization of antigen-specific IL-4- and IL-13-producing basophils and T cells in peripheral blood of healthy and asthmatic subjects. *J Allergy Clin Immunol*, 1999, 104, 811-819.

[58] Mitre, E; Taylor, RT; Kubofcik, J; Nutman, TB. Parasite antigen-driven basophils are a major source of IL-4 in human filarial infections. *J Immunol*, 2004, 172, 2439-2445.

[59] Galli, SJ; Wedemeyer, J; Tsai, M. Analyzing the roles of mast cells and basophils in host defense and other biological responses. *Int J Hematol*, 2002, 75, 363-369.

[60] Bieneman, AP; Chichester, KL; Chen, YH; Schroeder, JT. Toll-like receptor 2 ligands activate human basophils for both IgE-dependent and IgE-independent secretion. *J Allergy Clin Immunol*, 2005, 115, 295-301.

[61] Phillips, C; Coward, WR; Pritchard, DI; Hewitt, CR. Basophils express a type 2 cytokine profile on exposure to proteases from helminths and house dust mites. *J Leukoc Biol*, 2003, 73, 165-171.

[62] Westphal E. Ubermastzellen, in: P. Ehrlich (Eds.) Ferbenanlytische Untersuchungen Zur Histologie Und Klinik Des Blutes, Hirschwald, Berlin, 1891.

[63] Maltby, S; Khazaie, K; Nagny, KM. Mast cells in tumor growth: Angiogenesis, tissue remodelling and immune-modulation. *Biochim Biophys Acta*, 2009. [Epub ahead of print].

[64] Prager, MD; Bearden, J. Mast cell increase in mouse leukaemia. *Nature*, 1962, 193, 180-181.

[65] Shimizu, H; Nagakui, Y; Tsuchiya, K; Horii, Y. Demonstration of chymotryptic and tryptic activities in mast cells of rodents: comparison of 17 species of the family Muridae. *J Comp Pathol*, 2001, 125, 76-79.

[66] Grimbaldeston, MA; Finlay-Jones, JJ; Hart, PH. Mast cells in photodamaged skin: what is their role in skin cancer? *Photochem Photobiol Sci*, 2006, 5, 177-183.

[67] Wasiuk, A; de Vries, VC; Hartmann, K; Roers, A; Noelle, RJ. Mast cells as regulators of adaptive immunity to tumours. *Clin Exp Immunol*, 2009, 155, 140-146.

[68] Ch'ng, S; Wallis, RA; Yuan, L; Davis, PF; Tan, ST. Mast cells and cutaneous malignancies. *Mod. Pathol*, 2006, 19, 149–159.

[69] Amini, RM; Aaltonen, K; Nevanlinna, H; Carvalho, R; Salonen, L; Heikkila, P; Blomqvist, C. Mast cells and eosinophils in invasive breast carcinoma. *BMC Cancer*, 2007, 7, 165.
[70] Gulubova, M; Vlaykova, T. Prognostic significance of mast cell number and microvascular density for the survival of patients with primary colorectal cancer. *J Gastroenterol Hepatol*, 2007. [Epub ahead of print]
[71] Munn, DH; Mellor, AL. The tumor-draining lymph node as an immune-privileged site. *Immunol Rev*, 2006, 213, 146-158.
[72] Gajewski, TF; Meng, Y; Blank, C; Brown, I; Kacha, A; Kline, J; Harlin, H. Immune resistance orchestrated by the tumor microenvironment. *Immunol Rev*, 2006, 213, 131-145.
[73] Crivellato, E; Nico, B; Ribatti, D. Mast cells and tumour angiogenesis: new insight from experimental carcinogenesis. *Cancer Lett*, 2008, 269, 1-6.
[74] Galinsky, DS; Nechushtan, H. Mast cells and cancer-no longer just basic science. *Crit Rev Oncol Hematol*, 2008, 68, 115-130.
[75] Grizzi, F; Franceschini, B; Gagliano, N; Barbieri, B; Arosio, B; Annoni, G; Chiriva-Internati, M; Dioguardi, N. Mast cell density: a quantitative index of liver acute inflammation. *Anal Quant Cytol Histol*, 2002, 24, 63-69.
[76] Lampiasi, N; Azzolina, A; Montalto, G; Cervello, M. Histamine and spontaneously released mast cell granules affect the cell growth of human hepatocellular carcinoma cells. *Exp Mol Med*, 2007, 39, 284-294.
[77] Peng, SH; Deng, H; Yang, JF; Xie, PP; Li, C; Li, H; Feng, DY. Significance and relationship between infiltrating inflammatory cell and tumor angiogenesis in hepatocellular carcinoma tissues. *World J Gastroenterol*, 2005, 11, 6521-6524.
[78] Jensen-Jarolim, E; Achatz, G; Turner, MC; Karagiannis, S; Legrand, F; Capron, M; Penichet, ML; Rodríguez, JA; Siccardi, AG; Vangelista, L; Riemer, AB; Gould, H. AllergoOncology: the role of IgE-mediated allergy in cancer. *Allergy*, 2008, 63, 1255-1266.
[79] Knittelfelder, R; Riemer, AB; Jensen-Jarolim, E. Mimotope vaccination - from allergy to cancer. *Expert Opin Biol Ther*, 2009, 9, 493-506.

In: Basophil Granulocytes
Editor: Paul K. Vellis
ISBN: 978-1-60741-797-2

Chapter 3

BASOPHILIC GRANULOCYTE: A REVIEW OF ITS CHARACTERISTICS AND ROLES

Toshihisa Tsuruta* and Kenzaburo Tani†

* Department of Internal Medicine, National Hospital Organization Kumamoto Medical Center, Japan

†Department of Medical Genomics, Medical Institute of Bioregulation. Kyushu University and Department of Advanced Cell and Molecular Therapy, Kyushu, University Hospital, Japan

ABSTRACT

Basophilic granulocytes (basophils) are a very small population of peripheral blood leukocytes. Because basophils have high-affinity immunoglobulin E receptors (FcεRI) and secrete chemical mediators that contain histamine, they are thought to be very similar to mast cells. However, research characterizing the function of basophils was slow to proceed and their unique function and importance were not established for a significant period of time. Recently, a series of studies have characterized the role of basophils in anaphylactic shock, chronic allergic reactions, and other human immunological reactions. These studies have shown that basophils are not a supplementary cell type but key players in very serious immune reactions. In this chapter we introduce the recently discovered characteristics of basophils. In addition, we consider how these aspects are clinically important and are connected with new cellular or molecular treatments for allergic reactions.

INTRODUCTION

Basophilic granulocytes (basophils) are a small population of peripheral blood leukocytes that were first described in 1879 by Ehrlich[1]. These cells contain unique cytoplasmic granules that stain with basophilic dyes, such as toluidine blue. The percentage of basophils in the peripheral blood is low (<1%), and they share physicochemical properties with other blood cells. Like mast cells, basophils possess high-affinity immunoglobulin (Ig) E receptors (FcεRI) that are cross-linked when the receptor bound IgE is engaged with the corresponding antigen ("allergen"). Receptor cross-linking results in the release of a number of mediators, which contain some elements that are common to both cell types. Based on their similarity to mast cells, basophils have been thought to play a minor and possibly redundant role as "circulating mast cells"[2, 3].

Not so long ago, some investigators thought that mice entirely lacked basophils because they could not be detected by normal hematologic staining (eg.Wright-Giemsa). In 1981, Urbina et al. showed that mouse basophils have a distinct morphology that is different from other species, which may have contributed to their inability to be detected in routine skin preparations[4]. Furthermore, the ultrastructural features of mouse basophils have been well defined[5], and many studies have been conducted on this unique cell population[6, 7].

Phenotype, Development, and Activation of Basophils

The early stages of basophil maturation and their relationship to other cell lineages are not well understood[2]. Basophils express a variety of cytokine receptors (IL-1RII (CD121b), IL-2Rα (CD25), IL-3Rα (CD123), IL-4Rα (CD124), IL-8R (CD128), GM-CSFRα (CD116), IFNRγ (CD119)), chemokine receptors [CCR1, CCR2, CCR3, CCR5 (CD195), CXCR1 (IL-8Rα), CXCR2 (IL-8Rβ), CXCR4 (CD170), CRTH2], complement receptors [CD11b (iC3bR), CD11c (C3biR), CD21 (C3dR), CD35 (C3bR, C4bR), CD45 (C3bR, C4bR), CD55 (C4b/2aR, C3b/BbR), CD59 (C5b-8R, C5b-9R)], homing receptors and related molecules [CD15s, CD62L, CD162, CD11a (LFA-1), CD18, CD29, CD44 (Pgp-1), CD49a (VLA-1), CD49d (VLA-4)], prostaglandin receptors, and Ig Fc receptors [CDw32 (FcγRIIA and B), FcεRI, FcγRIII[8-10].

Previous reports have suggested that basophils evolved from eosinophil/basophil progenitors, and this hypothesis is supported by the presence of granulocytes with hybrid eosinophil/basophil phenotypes in patients with chronic or acute myelogenous leukemia and in cell culture[11-13]. On the other hand, the possibility that mast cells and basophils share a common lineage arises from the observation that basophils with phenotypic features that are characteristic of mast cells can be found in patients with asthma, allergies, or allergic drug reactions[14]. Therefore, the current predominant model that mast cells and basophils originate from separate lineages is still debated and may have to be revised[2].

Our understanding of basophils has been advanced by the development of basophil-specific monoclonal antibodies, Bsp-1, 2D7, BB1 and 212H6[10, 15-17]. In addition, the 97A6 monoclonal antibody has been described as an antibody specific for mature mast cells, basophils, and their progenitors[18]. 97A6 does not react with any other hematopoietic or nonhematopoietic cell types. The epitope recognized by 97A6 may therefore be associated with the commitment of the CD34 precursor to a mast cell or basophil lineage that is distinct from other lineages[2].

Among the many cytokines that stimulate basophils, IL-3 is thought to be the main growth and differentiation factor for basophils[19, 20]. SCF together with IL-3 expand the progenitor pool of most hematopoietic cell types in the bone marrow, including mast cells and basophils[20].

As with mast cells, basophils express complete and functional FcεRI receptors, and cross-linking of these receptors leads to basophil activation, granule exocytosis and mediator release[21]. C3a and C5a can also activate basophils through the C3aR and C5aR complement receptors, respectively. Activation through any of these receptors leads to histamine release, eicosanoid synthesis, and IL-4 and IL-13 gene expression. In addition, basophil activation may be associated with increased CD18 and CD63 expression and decreased Leu-8 (CD62L) expression[8].

Basophils and Anaphylaxis

Anaphylaxis is an immune reaction that is induced upon exposure to food, wasp toxins, and allergens, such as medicine and latex, and can lead to a severe generalized allergy[22, 23]. Because anaphylaxis is often associated with rapid skin urticaria, decreased blood pressure, dyspnea, consciousness disorder, etc., this condition is dangerous and sometimes fatal. Therefore, anaphylaxis is a pathosis that concerns medical professionals and both rapid diagnosis and treatment are necessary. Previous work has shown that both mast cells and basophils perform important functions in anaphylaxis however, more recently it became clear that mast cells and basophils have quite different mechanisms and chemical mediators (Figure 1)[23].

(1) Classical pathway of anaphylaxis by mast cells and IgE

Mast cells are prevalent in the skin, mucous membrane, and the circumvascular, and they become sensitized to an allergen when allergen-IgE complexes bind the high-affinity IgE receptor FcεRI expressed on their cell surface. When re-exposed to the same allergen, the allergen cross-links the IgE/FcεRI complex on the surface of the mast cell, and an activation signal is transmitted that induces mast cell degranulation and the extracellular release of secretory granules that contain histamine and other chemical mediators[24, 25]. As a result, vascular permeability increases, bronchus smooth muscles contract, etc., leading to the characteristic anaphylactic symptoms such as rapid hypotension and dyspnea[26].

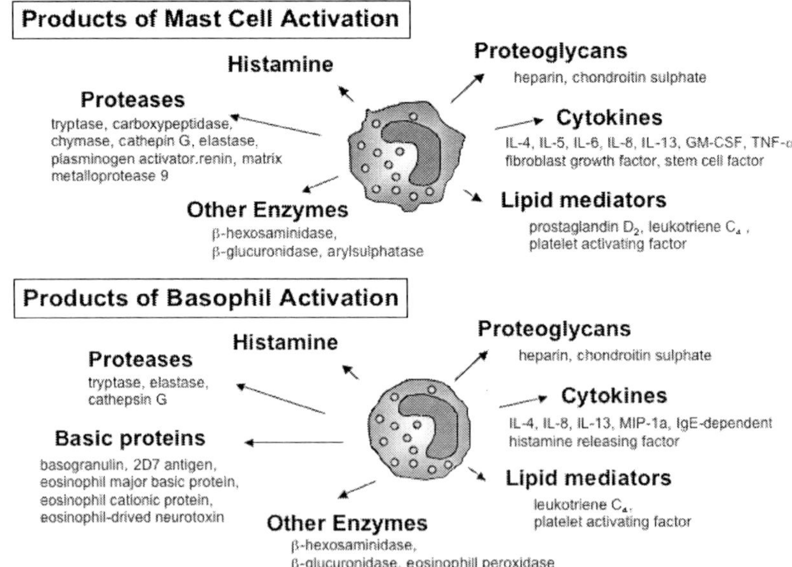

Figure 1. A. Mast cells and their activation products. B. Basophils and their activation products. Figure courtesy of Dr. A.F. Walls (Reference 23). IL, Interleukin; GM-CSF, Granulocyte-macrophage colony stimulating factor; TNF, Tumor necrosis factor; MIP, Macrophage inflammatory protein.

It was recently shown that anaphylactic pathosis cannot be completely attributed to the classical pathway based on analyses of a mouse model of the 'classical' pathway, which consists of IgE, FcεRI, mast cells, and histamine. For example, mast cell-deficient, IgE-deficient, or FcεRI-deficient mice still exhibited generalized anaphylaxis when they were sensitized to an allergen and then re-exposed to the same allergen[27-31]. These findings strongly suggest that there is another anaphylactic route that is distinct from the classical pathway (Figure 2)[32].

(2) A new anaphylaxis pathway that involves basophils and IgG

Mice lacking the FcRγ chain common to FcεRI, which is the IgE receptor, and FcγRI/III, which is the IgG receptor, do not experience anaphylactic reactions, suggesting that there is an IgG-mediated anaphylactic response in addition to an IgE-mediated reaction. After these mice were given antigen-specific IgG_1 monoclonal antibodies and then challenged with intravenous allergen, systemic anaphylaxis with tachycardia was induced[29]. Furthermore, it was ascertained that IgG_1-mediated anaphylaxis is induced in mast cell-deficient mice, which strongly suggest that a type of IgG_1-mediated, mast cell-independent anaphylaxis exists[27, 29]. Macrophages were considered a potential cell type that is responsible for IgG_1-mediated anaphylaxis[31], but there has been no conclusive proof establishing their role in this response. Because the high-affinity IgE receptor, FcεRI, is expressed on both mast cells and basophils in mice, it is conceivable that basophils are also are responsible for IgE-mediated anaphylaxis[33].

Tsujimura et al. developed a mouse monoclonal antibody that can deplete basophils in vivo. Using this antibody, they showed that basophils were dispensable for IgE-mediated anaphylaxis, while mast cells were critical for the "classical" anaphylactic pathway mediated by IgE and histamine. They injected allergen-specific IgE into basophil-depleted mice, and then challenged these mice with allergen. The anaphylactic response occurred in the basophil-depleted mouse, while mast cell-deficient mice exhibited no anaphylactic responses[34].

They also developed a penicillin G–specific mouse monoclonal IgG_1 antibody, and injected this IgG1 antibody into basophil-depleted mice. While an anaphylactic reaction occurred in control mice that were injected with penicillin G combined with bovine serum albumin, basophil-depleted mice did not have an anaphylactic response. Interestingly, this IgG-dependent anaphylactic response was more severe than the IgE-dependent response. Furthermore, these responses occurred in both normal and mast cell-deficient mice. Collectively, these experiments provided seminal evidence that basophils play an important role in IgG-dependent anaphylaxis [34]. The allergen-specific IgG_1 and allergen complexes were found mainly on the surface of basophils, but also on macrophages, neutrophils, and natural killer cells by flow cytometry analysis. However, if these cell types except basophils were depleted from the mice prior to injecting the antibody and allergen, there was still an IgG-dependent anaphylactic response[34]. These results strongly indicate that basophils are a key component of IgG-dependent anaphylaxis.

(3) Platelet-activating factor and basophil-mediated anaphylaxis

Unlike IgG-dependent anaphylaxis, IgE-dependent anaphylaxis in mice is prevented by pre-treating with an anti-histamine drug. Based on these findings, the intervention of chemical mediators other than histamine likely contributes to IgG-dependent anaphylaxis[34]. Platelet-activating factor (PAF) is also known to act on the vascular endothelium and facilitate its permeability. When mice were injected with a PAF antagonist, IgG-dependent, but not IgE-dependent, anaphylaxis was almost completely inhibited. The stimulation by allergen and allergen-specific IgG_1 caused a significant elevation in PAF production, only in basophils. Furthermore, when conditioned medium from basophils that were stimulated with immune complexes was added to human vascular endothelial cells (normal human umbilical vein endothelium cells: HUVEC), their intercellular space expanded as an indication of vascular permeability; furthermore, this phenotype was inhibited with a PAF antagonist. From these results, it became clear that basophils quickly combine the immune complexes that are formed in the blood through the IgG receptor, which induces the production and release of factors such as PAF that ultimately lead to systemic anaphylaxis[34]. Although basophils are a minor population of cells in the blood and occupy only 0.5% of peripheral leukocytes, when they are activated by immune complexes they induce strong systemic anaphylaxis by secreting PAF, which induces vascular permeability 1,000-10,000 times higher than histamine.

(4) The role of basophils in human anaphylaxis

The experiments described above analyzed passive anaphylaxis, which is induced by challenging with an allergen after administering allergen-specific IgE or IgG. To examine anaphylaxis reactions that are more similar to human allergic conditions, Tsujimura et al. examined the role of basophils in active anaphylaxis by immunizing mice with an allergen two weeks prior to an intravenous injection of the allergen. This active anaphylaxis is more serious than passive anaphylaxis and caused fatal anaphylactic shock not only in normal mice but also in mast cell-deficient mice. However, death from anaphylactic shock was prevented in mast cell-deficient mice when they were pretreated with the basophil-depleting antibody. This finding provided clear evidence that basophils have a decisive role in active anaphylaxis. Interestingly, fatal anaphylactic shock was not prevented in basophil-depleted normal mice that have mast cells. Therefore, these results indicate that both the classical pathway caused by mast cells and the new pathway caused by basophils equally contribute to anaphylactic shock[34].

Although it is not clear whether these results in mice apply to humans, there are many human reports that suggest there is an alternative pathway in addition to the classical pathway. There are reports of clinical anaphylactic cases, especially due to drug allergies, where allergen-specific IgE was not detected[35], mast cell tryptase was not elevated[36], and allergen-specific IgG antibody was elevated[37-39]. Recently, Vadas P. et al. showed that serum PAF levels were directly correlated and serum PAF acetylhydrolase activity was inversely correlated with the severity of anaphylaxis, and that the failure of PAF acetylhydrolase to inactivate PAF may contribute to the severity of anaphylaxis[40].

Based on analyses in mice, the new basophil-mediated anaphylaxis pathway is thought to require more allergens and antibodies than the classical pathway[34]. Therefore, when large amounts of soluble material are introduced into the body, such as antibody therapy, which has received recent attention as a molecular therapy, both the classical pathway and the new pathway may contribute to intense anaphylactic shock. Moreover, when we use peptide-specific immune therapy for autoimmune diseases or hyposensitization treatment for allergic diseases to induce immunologic tolerance, there is a danger that these therapies will lead to IgG-mediated anaphylaxis. Therefore, high-risk patients should be monitored not only for IgE levels but also for allergen-specific IgG, basophil function and/or serum PAF levels.

Roles of Basophils in Chronic Allergic Reactions

Basophils are often recruited to the site of allergic inflammation. However, for a long period of time, there was no definitive evidence that basophils were crucially involved in the pathogenesis of chronic allergic disorders. Mukai et al. showed that basophils are responsible for the development of IgE mediated chronic allergic inflammation independently of T cells and mast cells[41]. Using a chronic cutaneous allergy mouse model they showed that basophils act as an initiator cell rather than an effector cell in chronic allergic inflammation.

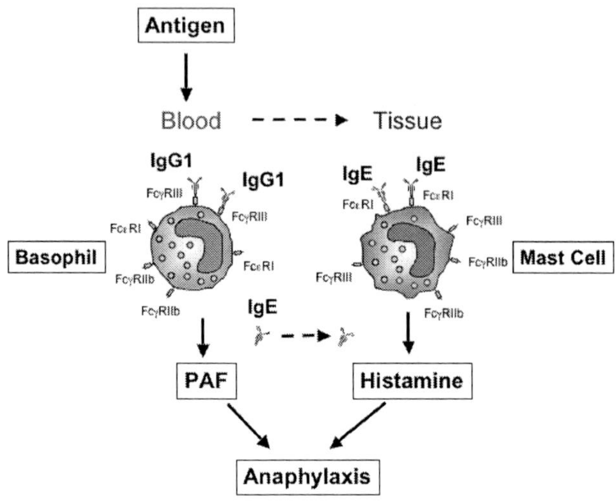

Figure 2. Schematic of the two anaphylaxis pathways. The classical pathway involving mast cells is initiated when antigen-IgE immune complexes bind FcεRI on the cell surface. The anaphylactic reaction is caused by the release of histamine from mast cells. The new pathway involving basophils begins when antigen-IgG_1 immune complexes bind to FcγRIII surface receptors, and platelet-activating factor (PAF) is released from basophils.

(1) Roles of IgE in the chronic allergic reaction

Matsuoka et al. established transgenic mice that carry genes encoding the heavy and light chains of hapten trinitrophenol (TNP)-specific IgE to generate a model system that would help elucidate both the pathological roles of IgE in the acute and chronic phases of allergic inflammation and the immunobiological roles in vivo. These mice produced high titers of TNP-specific IgE and their mast cells were heavily loaded with IgE[42]. According to their colleague's report, the immediate-type allergic ear swelling response of the biphasic response appeared after the ear of the mouse was intracutaneously challenged with the homologous TNP-binding ovalbumin[43]. Furthermore, on the second day of the antigenic challenge the ear began to swell again and more than doubled in skin thickness on the fourth day. Pathologic histology was used to show the detailed aspects of chronic allergic inflammation; for example, there was a robust infiltration of cells including basophils, and cornification was recognized as hyperplasia of the epidermis. The third phase of ear swelling was shown to be antigen-specific and IgE-dependent and it could be induced when antigen-specific IgE was given to normal mice (passive sensitization) prior to the antigen challenge. This phenomenon was seen both in transgenic and wild-type mice. It was proven that IgE contributed not only to the immediate-type allergic response but also to the chronic allergic inflammatory response[41].

(2) The role of basophils in the IgE-dependent chronic allergic inflammatory response

When mast cell-deficient mice were challenged with the antigen after passive sensitization of IgE, the first and second phases of ear swelling, which are the immediate-type allergy response, were not observed, while the third-phase of ear swelling was observed both in these mice and in normal mice[41]. Based on this phenomenon, it is thought that mast cells are dispensable for the IgE-dependent chronic allergic response and that the immediate-type allergic response is required to induce the chronic response. Cyclosporine A almost completely inhibited the third phase of ear swelling and cellular infiltration, whereas an anti-histamine, cyproheptadine, did not have any significant effects on the third phase of the reaction. Given the delayed time of ear swelling, T cells were thought to contribute to this response. However, the third phase of ear swelling was also observed in T cell-deficient mice and T cells were not essential for the initiation of IgE-dependent chronic allergic inflammation[43].

Neither immediate ear swelling nor the third phase of ear swelling was observed in mice deficient for FcεRI, which is a high-affinity IgE receptor. This result indicates that the third phase of ear swelling requires a cell type that expresses FcεRI on its surface. To identify this cell type, Mukai et al. transferred various cells from normal mice into FcεRI-deficient mice and studied whether the third phase of ear swelling was restored. When FcεRI-expressing basophils that also express the natural killer cell marker DX5 (CD49b) were transferred into FcεRI-deficient mice, the third phase of ear swelling was restored. These findings indicate that this novel mechanism that leads to the development of chronic allergic inflammation is induced by basophils through the interaction of antigen, IgE, and FcεRI[41].

(3) Basophils as potential therapeutic targets for chronic allergic inflammation

When Mukai et al. examined the cells that infiltrated into the skin during the third phase of ear swelling skin, basophils comprised only 1-2% of the infiltrate and most of the cells were eosinophils and neutrophils. Therefore, they wanted to determine how this minor basophil population could cause chronic allergic inflammation[41].

Using the basophil-depleting monoclonal antibody established by their colleagues, they confirmed that this antibody could markedly decrease the number of basophils and further showed that mice pretreated with this antibody did not exhibit the third phase of ear swelling[44, 45]. This result proves that basophils are responsible for IgE-dependent chronic allergic inflammation. Furthermore, when this basophil-depleting antibody was given 2 or 3 days after allergen injection and the third-phase ear swelling had already occurred, the ear swelling and inflammation were inhibited and eosinophil and neutrophil infiltration was decreased markedly[44]. This result suggests that basophils function more as an initiator cell than as an effector cell and indicates the possibility that chronic allergic inflammation could be treated by targeting basophils. Once activated by allergen-mediated cross-linking of IgE/FcεRI, basophils secrete humoral factors such as a cytokines and chemokines that may directly or indirectly contribute to the infiltration of eosinophils and neutrophils.

The Role of Basophils in the Control of T Cell Differentiation

(1) T cell development and IL-4

CD4 positive T cells are functionally divided into four types; Th1, Th2, and Th17 are helper T cells[46-49], while Tregs are regulatory T cells[50]. Naive T cells differentiate into these functional T cells in response to different cytokines. IL-4 plays a crucial role in the development of Th2-type immune responses and the regulation of immunoglobulin isotype switching to IgE[51-53]. The IL-4-producing cell that produces sufficient levels of IL-4 to differentiate naive cells into Th2 cells in the lymph node is still unknown. Because Th2 cells produce variable amounts of Il-4, T cells themselves may be the predominant IL-4-producing cell that stimulates naive T cells to differentiate into Th2 cells[51, 54]. On the other hand, dendritic cells (DCs) and macrophages are required for CD4+ T cells to develop into Th1 cells. DCs are generally divided into three types; DC1 cells express the highest levels of MHC class I, class II, CD40, B7.1 and B7.2 compared to DC0 and DC2 cells. In terms of IL-12 production, DC1 cells have enhanced production, while DC2 cells produce lower levels than DC0 cells. Both DC0 and DC1 supported the differentiation of IFNγ-producing Th1 cells, but not IL-4-producing Th2 cells from TCR-transgenic naïve mouse Th cells. However, DC2 cells selectively enhanced the differentiation of IL-4-producing Th2 cells[55].

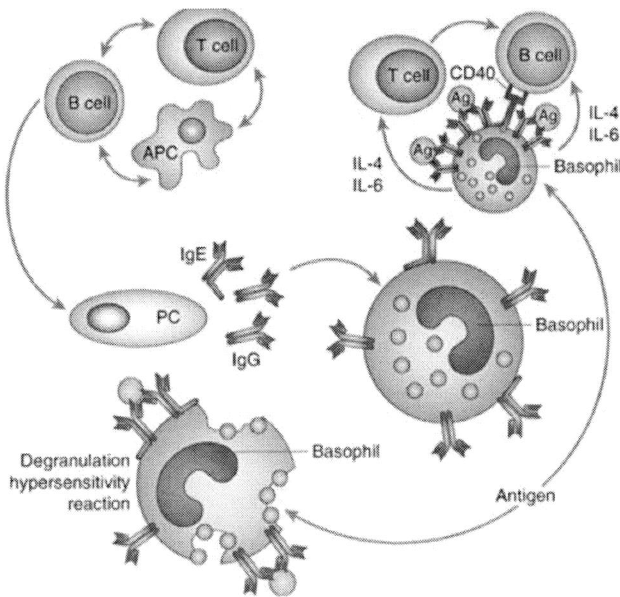

Figure 3. Basophils are mediators of the immune response. Upper left: Classical primary response by CD4+ T cells, B cells, and APCs, Ag-presenting cell (APC); PC, plasma cell. After IgG and IgE are produced by plasma cells (PC), basophils bind these circulating Abs. Upper right: A second presentation with Ag results in Ag-loaded basophils that begin producing the cytokines IL-4 and IL-6. Ag presentation and cytokine production by basophils enhance the secondary humoral immune response (upper right). The hypersensitivity reaction is caused by basophil degranulation and the release of lipid mediators (lower left). (Reference 64)

(2) The roles of basophils in Th2 cell development

Based on previous studies, basophils along with mast cells were thought to be cell types that produce chemical mediators such as the histamine and leukotoluene and cause allergic reactions. However, recently, basophils have received much attention because they have been shown to produce massive amounts of IL-4 when activated[56-59]. A basophil produces ten-fold more IL-4 than a Th2 cell. When naïve T-cells were cultivated with activated basophils in vitro, the naïve T cells differentiated into Th2 cells as a direct result of the IL-4 that was secreted by basophils[60, 61]. In addition, it has been established that parasitic infected mice have an abundance of differentiated Th2 cells. It was also reported that basophils produced IL-4 under such conditions[58, 59]. However, it is still unknown whether the IL-4 produced by basophils initiates Th2 cell differentiation or maintains the Th2 cell dominant status. Furthermore, it was also unknown whether basophils interact with the naïve T cells in vivo. Naïve T cells differentiate into Th2 cells in the peripheral lymph node, and it is thought that basophils, unlike lymphocytes, circulate in the peripheral blood and never enter the peripheral lymph nodes. However, Sokol et al. recently elucidated a novel role of basophils in immunized mice[62]. They noted that many allergens have protease activity; therefore, they injected the protease papain into mice and then examined the secretion of IL-4 and other Th2-inducing cytokines. They showed that basophils transiently entered the regional lymph nodes soon after papain administration and that basophils secreted IL-4 after direct stimulation with papain. These results suggest that basophils also have important roles in the initial stages of immunization.

Roles of Basophils in Immunological Memory Responses

The cellular basis of immunological memory remains controversial. The classical primary immune response by CD4+ T cells, B cells, and antigen–presenting cells (APC) is initiated when antigen is presented to the host. Following IgG and IgE production by plasma cells (PC), basophils bind circulating antibody of either class (Figure 3). Denzel et al. showed that basophils bound large amounts of intact antigens on their surface and were the main source of IL-4 and IL-6 in the spleen and bone marrow after restimulation with a soluble antigen[63]. They also showed that basophil depletion resulted in a much lower humoral memory response and greater susceptibility of immunized mice to *Streptococcus pneumoniae*-induced sepsis. Adoptive transfer of antigen-reactive basophils significantly increased specific antibody production, and activated basophils, together with CD4+ T cells, profoundly enhanced B cell proliferation and immunoglobulin production. These basophil-dependent effects on B cells required IL-4 and IL-6 and increased the capacity of CD4+ T cells to provide B cell help (Figure 3. upper right). The circumstances that differentiate this participatory role of basophils from its traditional mast cell-like reaction, consisting of full degranulation, lipid mediator release, and cytokine release and subsequent anaphylactic reaction, remain unknown (Figure 3)[64].

Although basophils are not an abundant cell type and have been previously overlooked, they have recently emerged as highly important immune cells. Because anaphylactic shock and other serious reactions that involve basophils are closely related and can cause serious conditions, it is possible that basophils are maintained at low numbers to avoid these conditions. However, because they are usually low in number, they may be easy to target for treatment. Because brilliant research in recent years has greatly contributed to our understanding of the importance of basophils, further research on the function of basophils will be more forthcoming in the future. As for the development of new treatments, basophils can now be viewed as essential components of immune responses and potential therapeutic targets.

References

[1] Ehrlich, P. Beiträge zur Kenntnis der granulierten Bindegeweszsllen und der eosinophilen Leukocythen. *Arch Anat Physiol.*, 1879, 166-9.

[2] Falcone, FH; Haas, H; Gibbs, BF. The human basophil: a new appreciation of its role in immune responses. *Blood.*, 2000 Dec 15, 96(13), 4028-38.

[3] Prussin, C; Metcalfe, DD. 4. IgE, mast cells, basophils, and eosinophils. *J Allergy Clin Immunol.*, 2003 Feb, 111(2 Suppl), S486-94.

[4] Urbina, C; Ortiz, C; Hurtado, I. A new look at basophils in mice. *Int Arch Allergy Appl Immunol.*, 1981, 66(2), 158-60.

[5] Dvorak, AM; Nabel, G; Pyne, K; Cantor, H; Dvorak, HF; Galli, SJ. Ultrastructural identification of the mouse basophil. *Blood.*, 1982 Jun, 59(6), 1279-85.

[6] Lee, JJ; McGarry, MP. When is a mouse basophil not a basophil? *Blood.*, 2007 Feb 1, 109(3), 859-61.

[7] Galli, SJ; Franco, CB. *Basophils are back! Immunity.*, 2008 Apr, 28(4), 495-7.

[8] Valent, P. The phenotype of human eosinophils, basophils, and mast cells. *J Allergy Clin Immunol.*, 1994 Dec, 94(6 Pt 2), 1177-83.

[9] Agis, H; Fureder, W; Bankl, HC; Kundi, M; Sperr, WR; Willheim, M; et al. Comparative immunophenotypic analysis of human mast cells, *blood basophils and monocytes. Immunology.*, 1996 Apr, 87(4), 535-43.

[10] Arock, M; Schneider, E; Boissan, M; Tricottet, V; Dy, M. Differentiation of human basophils: an overview of recent advances and pending questions. *J Leukoc Biol.*, 2002 Apr, 71(4), 557-64.

[11] Leary, AG; Ogawa, M. Identification of pure and mixed basophil colonies in culture of human peripheral blood and marrow cells. *Blood.*, 1984 Jul, 64(1), 78-83.

[12] Denburg, JA; Telizyn, S; Messner, H; Lim, B; Jamal, N; Ackerman, SJ. et al. Heterogeneity of human peripheral blood eosinophil-type colonies: evidence for a common basophil-eosinophil progenitor. *Blood.*, 1985 Aug, 66(2), 312-8.

[13] Weil, SC; Hrisinko, MA. A hybrid eosinophilic-basophilic granulocyte in chronic granulocytic leukemia. *Am J Clin Pathol.*, 1987 Jan, 87(1), 66-70.

[14] Li, L; Li, Y; Reddel, SW; Cherrian, M; Friend, DS; Stevens, RL. et al. Identification of basophilic cells that express mast cell granule proteases in the peripheral blood of asthma, allergy, and drug-reactive patients. *J Immunol.*, 1998 Nov 1, 161(9), 5079-86.

[15] Bodger, MP; Mounsey, GL; Nelson, J; Fitzgerald, PH. A monoclonal antibody reacting with human basophils. *Blood.*, 1987 May, 69(5), 1414-8.

[16] Kepley, CL; Craig, SS; Schwartz, LB. Identification and partial characterization of a unique marker for human basophils. *J Immunol.*, 1995 Jun 15, 154(12), 6548-55.

[17] McEuen, AR; Buckley, MG; Compton, SJ; Walls, AF. Development and characterization of a monoclonal antibody specific for human basophils and the identification of a unique secretory product of basophil activation. *Lab Invest.*, 1999 Jan, 79(1), 27-38.

[18] Buhring, HJ; Simmons, PJ; Pudney, M; Muller, R; Jarrossay, D; van Agthoven, A. et al. The monoclonal antibody 97A6 defines a novel surface antigen expressed on human basophils and their multipotent and unipotent progenitors. *Blood.*, 1999 Oct 1, 94(7), 2343-56.

[19] Kirshenbaum, AS; Goff, JP; Kessler, SW; Mican, JM; Zsebo, KM; Metcalfe, DD. Effect of IL-3 and stem cell factor on the appearance of human basophils and mast cells from CD34+ pluripotent progenitor cells. *J Immunol.*, 1992 Feb 1, 148(3), 772-7.

[20] Valent, P; Schmidt, G; Besemer, J; Mayer, P; Zenke, G; Liehl, E. et al. Interleukin-3 is a differentiation factor for human basophils. *Blood.*, 1989 May 15, 73(7), 1763-9.

[21] Kinet, JP. The high-affinity IgE receptor (Fc epsilon RI): from physiology to pathology. *Annu Rev Immunol.*, 1999, 17, 931-72.

[22] Sampson, HA; Munoz-Furlong, A; Campbell, RL; Adkinson, NF; Jr., Bock, SA; Branum, A. et al. Second symposium on the definition and management of anaphylaxis: summary report--Second National Institute of Allergy and Infectious Disease/Food Allergy and Anaphylaxis Network symposium. *J Allergy Clin Immunol.*, 2006 Feb, 117(2), 391-7.

[23] Simons, FE; Frew, AJ; Ansotegui, IJ; Bochner, BS; Golden, DB; Finkelman, FD. et al. Risk assessment in anaphylaxis: current and future approaches. *J Allergy Clin Immunol.*, 2007 Jul, 120(1 Suppl), S2-24.

[24] Turner, H; Kinet, JP. Signalling through the high-affinity IgE receptor Fc epsilonRI. *Nature.*, 1999 Nov 25, 402(6760 Suppl), B24-30.

[25] Kraft, S; Kinet, JP. New developments in FcepsilonRI regulation, function

and inhibition. *Nat Rev Immunol.*, 2007 May, 7(5), 365-78.

[26] Winbery, SL; Lieberman, PL. Histamine and antihistamines in anaphylaxis. *Clin Allergy Immunol.*, 2002, 17, 287-317.

[27] Jacoby, W; Cammarata, PV; Findlay, S; Pincus, SH. Anaphylaxis in mast cell-deficient mice. *J Invest Dermatol.*, 1984 Oct, 83(4), 302-4.

[28] Dombrowicz, D; Flamand, V; Miyajima, I; Ravetch, JV; Galli, SJ; Kinet, JP. Absence of Fc epsilonRI alpha chain results in upregulation of Fc gammaRIII-dependent mast cell degranulation and anaphylaxis. Evidence of competition between Fc epsilonRI and Fc gammaRIII for limiting amounts of FcR beta and gamma chains. *J Clin Invest.*, 1997 Mar 1, 99(5), 915-25.

[29] Miyajima, I; Dombrowicz, D; Martin, TR; Ravetch, JV; Kinet, JP; Galli, SJ. Systemic anaphylaxis in the mouse can be mediated largely through IgG1 and Fc gammaRIII. Assessment of the cardiopulmonary changes, mast cell degranulation, and death associated with active or IgE- or IgG1-dependent passive anaphylaxis. *J Clin Invest.*, 1997 Mar 1, 99(5), 901-14.

[30] Oettgen, HC; Martin, TR; Wynshaw-Boris, A; Deng, C; Drazen, JM; Leder, P. Active anaphylaxis in IgE-deficient mice. *Nature.*, 1994 Aug 4, 370(6488), 367-70.

[31] Strait, RT; Morris, SC; Yang, M; Qu, XW; Finkelman, FD. Pathways of anaphylaxis in the mouse. *J Allergy Clin Immunol.*, 2002 Apr, 109(4), 658-68.

[32] Finkelman, FD. Anaphylaxis: lessons from mouse models. *J Allergy Clin Immunol.*, 2007 Sep, 120(3), 506-15, quiz 16-7.

[33] Galli, SJ. Pathogenesis and management of anaphylaxis: current status and future challenges. *J Allergy Clin Immunol.*, 2005 Mar, 115(3), 571-4.

[34] Tsujimura, Y; Obata, K; Mukai, K; Shindou, H; Yoshida, M; Nishikado, H. et al. Basophils play a pivotal role in immunoglobulin-G-mediated but not immunoglobulin-E-mediated systemic anaphylaxis. *Immunity.*, 2008 Apr, 28(4), 581-9.

[35] Cheifetz, A; Smedley, M; Martin, S; Reiter, M; Leone, G; Mayer, L. et al. The incidence and management of infusion reactions to infliximab: a large center experience. *Am J Gastroenterol.*, 2003 Jun, 98(6), 1315-24.

[36] Dybendal, T; Guttormsen, AB; Elsayed, S; Askeland, B; Harboe, T; Florvaag, E. Screening for mast cell tryptase and serum IgE antibodies in 18 patients with anaphylactic shock during general anaesthesia. *Acta Anaesthesiol Scand.*, 2003 Nov, 47(10), 1211-8.

[37] Kraft, D; Hedin, H; Richter, W; Scheiner, O; Rumpold, H; Devey, ME. Immunoglobulin class and subclass distribution of dextran-reactive

antibodies in human reactors and non reactors to clinical dextran. *Allergy.*, 1982 Oct, 37(7), 481-9.

[38] Adourian, U; Shampaine, EL; Hirshman, CA; Fuchs, E; Adkinson, NF Jr. High-titer protamine-specific IgG antibody associated with anaphylaxis: report of a case and quantitative analysis of antibody in vasectomized men. *Anesthesiology.*, 1993 Feb, 78(2), 368-72.

[39] Weiss, ME; Nyhan, D; Peng, ZK; Horrow, JC; Lowenstein, E; Hirshman, C. et al. Association of protamine IgE and IgG antibodies with life-threatening reactions to intravenous protamine. *N Engl J Med.*, 1989 Apr 6, 320(14), 886-92.

[40] Vadas, P; Gold, M; Perelman, B; Liss, GM; Lack, G; Blyth, T. et al. Platelet-activating factor, PAF acetylhydrolase, and severe anaphylaxis. *N Engl J Med.*, 2008 Jan 3, 358(1), 28-35.

[41] Mukai, K; Matsuoka, K; Taya, C; Suzuki, H; Yokozeki, H; Nishioka, K. et al. Basophils play a critical role in the development of IgE-mediated chronic allergic inflammation independently of T cells and mast cells. *Immunity.*, 2005 Aug, 23(2), 191-202.

[42] Matsuoka, K; Taya, C; Kubo, S; Toyama-Sorimachi, N; Kitamura, F; Ra, C. et al. Establishment of antigen-specific IgE transgenic mice to study pathological and immunobiological roles of IgE in vivo. *Int Immunol.*, 1999 Jun, 11(6), 987-94.

[43] Sato, E; Hirahara, K; Wada, Y; Yoshitomi, T; Azuma, T; Matsuoka, K. et al. Chronic inflammation of the skin can be induced in IgE transgenic mice by means of a single challenge of multivalent antigen. *J Allergy Clin Immunol.*, 2003 Jan, 111(1), 143-8.

[44] Obata, K; Mukai, K; Tsujimura, Y; Ishiwata, K; Kawano, Y; Minegishi, Y. et al. Basophils are essential initiators of a novel type of chronic allergic inflammation. *Blood.*, 2007 Aug 1, 110(3), 913-20.

[45] Kojima, T; Obata, K; Mukai, K; Sato, S; Takai, T; Minegishi, Y. et al. Mast cells and basophils are selectively activated in vitro and in vivo through CD200R3 in an IgE-independent manner. *J Immunol.*, 2007 Nov 15, 179(10), 7093-100.

[46] Mosmann, TR; Coffman, RL. TH1 and TH2 cells: different patterns of lymphokine secretion lead to different functional properties. *Annu Rev Immunol.*, 1989, 7, 145-73.

[47] Infante-Duarte, C; Horton, HF; Byrne, MC; Kamradt, T. Microbial lipopeptides induce the production of IL-17 in Th cells. *J Immunol.*, 2000 Dec 1, 165(11), 6107-15.

[48] Harrington, LE; Hatton, RD; Mangan, PR; Turner, H; Murphy, TL;

Murphy, KM. et al. Interleukin 17-producing CD4+ effector T cells develop via a lineage distinct from the T helper type 1 and 2 lineages. *Nat Immunol.*, 2005 Nov, 6(11), 1123-32.

[49] Park, H; Li, Z; Yang, XO; Chang, SH; Nurieva, R; Wang, YH. et al. A distinct lineage of CD4 T cells regulates tissue inflammation by producing interleukin 17. *Nat Immunol.*, 2005 Nov, 6(11), 1133-41.

[50] McGuirk, P; Mills, KH. Pathogen-specific regulatory T cells provoke a shift in the Th1/Th2 paradigm in immunity to infectious diseases. *Trends Immunol.*, 2002 Sep, 23(9), 450-5.

[51] Seder, RA; Paul, WE. Acquisition of lymphokine-producing phenotype by CD4+ T cells. *Annu Rev Immunol.*, 1994, 12, 635-73.

[52] Finkelman, FD; Holmes, J; Katona, IM; Urban, JF. Jr., Beckmann, MP; Park, LS. et al. Lymphokine control of in vivo immunoglobulin isotype selection. *Annu Rev Immunol.*, 1990, 8, 303-33.

[53] Mowen, KA; Glimcher, LH. Signaling pathways in Th2 development. *Immunol Rev.*, 2004 Dec, 202, 203-22.

[54] Guo, L; Hu-Li, J; Zhu, J; Watson, CJ; Difilippantonio, MJ; Pannetier, C. et al. In TH2 cells the Il4 gene has a series of accessibility states associated with distinctive probabilities of IL-4 production. *Proc Natl Acad Sci*, U S A. 2002 Aug 6, 99(16), 10623-8.

[55] Sato, M; Iwakabe, K; Ohta, A; Sekimoto, M; Nakui, M; Koda, T. et al. Functional heterogeneity among bone marrow-derived dendritic cells conditioned by T(h)1- and T(h)2-biasing cytokines for the generation of allogeneic cytotoxic T lymphocytes. *Int Immunol.*, 2000 Mar, 12(3), 335-42.

[56] Seder, RA; Paul, WE; Dvorak, AM; Sharkis, SJ; Kagey-Sobotka, A; Niv, Y. et al. Mouse splenic and bone marrow cell populations that express high-affinity Fc epsilon receptors and produce interleukin 4 are highly enriched in basophils. *Proc Natl Acad Sci*, U S A. 1991 Apr 1, 88(7), 2835-9.

[57] Schroeder, JT; MacGlashan, DW; Jr., Lichtenstein, LM. Human basophils: mediator release and cytokine production. *Adv Immunol.*, 2001, 77, 93-122.

[58] Voehringer, D; Shinkai, K; Locksley, RM. Type 2 immunity reflects orchestrated recruitment of cells committed to IL-4 production. *Immunity.*, 2004 Mar, 20(3), 267-77.

[59] Min, B; Prout, M; Hu-Li, J; Zhu, J; Jankovic, D; Morgan, ES. et al. Basophils produce IL-4 and accumulate in tissues after infection with a Th2-inducing parasite. *J Exp Med.*, 2004 Aug 16, 200(4), 507-17.

[60] Hida, S; Tadachi, M; Saito, T; Taki, S. Negative control of basophil expansion by IRF-2 critical for the regulation of Th1/Th2 balance. *Blood.*, 2005 Sep 15, 106(6), 2011-7.

[61] Oh, K; Shen, T; Le Gros, G; Min, B. Induction of Th2 type immunity in a mouse system reveals a novel immunoregulatory role of basophils. *Blood.*, 2007 Apr 1, 109(7), 2921-7.

[62] Sokol, CL; Barton, GM; Farr, AG; Medzhitov, R. A mechanism for the initiation of allergen-induced T helper type 2 responses. *Nat Immunol.*, 2008 Mar, 9(3), 310-8.

[63] Denzel, A; Maus, UA; Rodriguez Gomez, M; Moll, C; Niedermeier, M; Winter, C; et al. Basophils enhance immunological memory responses. *Nat Immunol.*, 2008 Jul, 9(7), 733-42.

[64] MacGlashan, D Jr. Granulocytes: new roles for basophils. *Immunol Cell Biol.*, 2008 Nov-Dec, 86(8), 637-8.

In: Basophil Granulocytes
Editor: Paul K. Vellis

ISBN: 978-1-60741-797-2

Chapter 4

FLAVONOIDS, NATURAL INHIBITORS OF BASOPHIL ACTIVATION

Toshio Tanaka**, Toru Hirano, Mari Kawai, Junsuke Arimitsu, Keisuke Hagihara, Masako Ogawa, Yusuke Kuwahara, Yoshihito Shima, Masashi Narazaki, Atsushi Ogata and Ichiro Kawase

Department of Respiratory Medicine, Allergy and Rheumatic Diseases, Osaka, University Graduate School of Medicine 2-2 Yamada-oka, Suita City, Osaka 565-0871, Japan.

ABSTRACT

Basophils comprise less than 1% of blood circulating granulocytes and share several similarities with mast cells, including expression of high affinity IgE receptor and the ability to secrete chemical mediators and cytokines. Basophils also express CD40 ligand, which in combination with IL-4 and IL-13 promotes IgE class switching in B cells. Flavonoids, ubiquitously present in vegetables, fruits or teas possess inhibitory activity of basophil activation. Flavonoids inhibit histamine release, synthesis of IL-4 and IL-13 and CD40 ligand expression by activated basophils. Among

* Corresponding author: Department of Respiratory Medicine, Allergy and Rheumatic Diseases, Osaka University Graduate School of Medicine, 2-2 Yamada-oka, Suita City, Osaka 565-0871, Japan; Tel: +81-6-6879-3838; Fax: +81-6-6879-3839; E-mail: ttanak@imed3.med.osaka-u.ac.jp

flavones, luteolin, apigenin and fisetin are the strongest inhibitors of IL-4 production with an IC_{50} value of 2-6 μM, while quercetin and kaempferol exert a substantial inhibitory activity with an IC_{50} value of 15-19 μM. The inhibitory activity of flavonoids against IL-4 and CD40 ligand expression is thought to be mediated through their inhibition action of activation of the nuclear factor of activated T cells and AP-1. These suppressive effects of flavonoids on basophil activation suggest that an appropriate daily intake of flavonoids might ameliorate allergic symptoms or prevent the onset of allergic diseases. In fact, administration of astragalin, a kaempferol glucoside, to atopic dermatitis-prone mice was found to prevent the onset of dermatitis and ameliorated the severity of dermatitis even after onset. In addition, intake of enzymatically modified isoquercitrin, a quercetin glycoside, had an ameliorative effect on ocular symptoms of patients with Japanese cedar pollinosis.

INTRODUCTION

Basophils account for less than 1% of blood circulating granulocytes and share several characteristics with mast cells, including expression of high affinity IgE receptors and the ability to secrete chemical mediators and cytokines [1, 2]. Basophils also express the CD40 ligand, which, in combination with IL-4 and IL-13, promotes IgE class switching in B cells [3]. Thus, basophils are believed to be actively involved in allergic responses. Indeed, a number of anti-allergic drugs in current use for the treatment of allergic diseases feature suppression of mediator release and cytokine expression by basophils [4]. Flavonoids, ubiquitous in fruits, vegetables and teas, are also inhibitors of basophil activation. This review article summarizes the anti-allergic activity of flavonoids and suggests that an appropriate intake of flavonoids may constitute a dietary means for amelioration and prevention of allergic diseases.

I. Flavonoids Inhibit IL-4, IL-13 and CD40 Ligand Expression by Basophils

In the mid-1990s we evaluated the clinical effect of one kind of traditional vegetarian diet on adult patients with atopic dermatitis. After a two-month treatment period, the severity of dermatitis decreased significantly with a reduction in lactate dehydrogenase-5 activity, the number of peripheral eosinophils and the amount of urinary secretion of 8-hydroxy-2'-deoxyguanosine, a marker of oxidative DNA damage [5, 6]. What factor(s) led to this amelioration of dermatitis remains unknown but subsequently it was found that one of the characteristics of the treatment was a high intake of flavonoids.

Flavonoids comprise of a large group of low-molecular-weight polyphenolic plant metabolites that are found in fruits, vegetables and teas, and thus are common substances in the daily diet [7, 8]. Flavonoids have been found to exert several biological activities including antioxidant, anti-bacterial and anti-viral activities, and to possess anti-inflammatory, anti-angionic, analgestic, hepatoprotective, cytostatic, apoptotic, estrogenic or anti-estrogenic as well as anti-allergic properties [9, 10]. Among the anti-allergic activities of flavonoids, Fewtress and Gomperts first identified the inhibition by flavones of transport ATPase in histamine secretion from rat mast cells [11], followed by the discovery of quercetin inhibition of allergen-stimulated human basophils [12, 13]. Apigenin, luteolin, 3.6-dihydroxy flavones, fisetin, kaempferol, quercetin, and myricetin were found to inhibit hexosaminidase release from rat mast cells with an IC_{50} value of less than 10 μM [14]. In addition flavonoids have also been shown to suppress cysteinyl leukotriene synthesis through inhibition of phospholipase $(PL)A_2$ and/ or 5-lipoxygenase (5LO) [15, 16]. As for the suppressive effect of flavonoids on cytokine expression, Kimata et al first reported that luteolin, quercetin and baicalein inhibited the secretion of granulocyte macrophage-colony stimulating factor by human cultured mast cells in response to cross-linkage of FcεRI [17] and subsequently showed that these compounds also inhibited IgE-mediated tumor necrosis factor (TNF)-α and IL-6 production by bone marrow-derived cultured murine mast cells [18]. These findings thus indicate that flavonoids are inhibitors of chemical mediator release and cytokine production by mast cells and perhaps basophils.

One of the characteristic features of allergic diseases is overproduction of IgE in response to environmental allergens. For the differentiation of B cells into IgE producing cells, both the interaction of the CD40 ligand with CD40 and the action of IL-4 or IL-13 on B cells are required [3]. The cells which are reportedly capable of sending these signals to B cells are Th2 cells, basophils and mast cells [19]. The effects of flavonoids on IL-4, IL-13 and CD40 ligand expression were therefore examined by using basophils. Peripheral blood basophils purified with the percoll gradient centrifugation and negative selection methods were stimulated with anti-IgE antibody or an allergen (Japanese cedar extract) in the presence or absence of various concentrations of flavonoids. Fisetin suppressed both IL-4 and IL-13 synthesis in a dose-dependent fashion by allergens or anti-IgE antibody-stimulated peripheral blood basophils and the IC_{50} value of fisetin for inhibition of IL-4 synthesis was 5.8 μM [20, 21]. Fisetin also inhibited IL-4, IL-5 and IL-13 production by KU812 cells, a basophilic cell line, in response to A23187 plus PMA, but the suppressive effect of fisetin on IL-6, IL-8 and IL-1β synthesis was relatively weak [20]. In order to determine the basic structure of flavonoids inhibiting IL-4 production and to identify more active compounds, 45 kinds of flavones, flavonols and their related compounds were screened [21, 22]. Luteolin, apigenin and fisetin were found to be the strongest inhibitors at an IC_{50} value of 2.7-5.8 μM. Quercetin and kaempferol are representative flavonoids with a substantial daily intake and had an intermediate inhibitory effect on IL-4 synthesis at an IC_{50} value of 15.7-18.8 μM, while myricetin showed no such activity (Figure 1). In addition, luteolin, apigenin and fisetin suppressed CD40 ligand expression by PMA+A23187-stimulated basophils and KU812 cells in a dose-dependent manner, whereas myricetin even at 30 μM did not show such an activity [23]. The hierarchy of the inhibitory effects of flavonoids on CD40 ligand expression was similar to those on IL-4 synthesis by basophils. Flavonoids such as luteolin, apigenin and fisetin should therefore be considered potential natural IgE inhibitors.

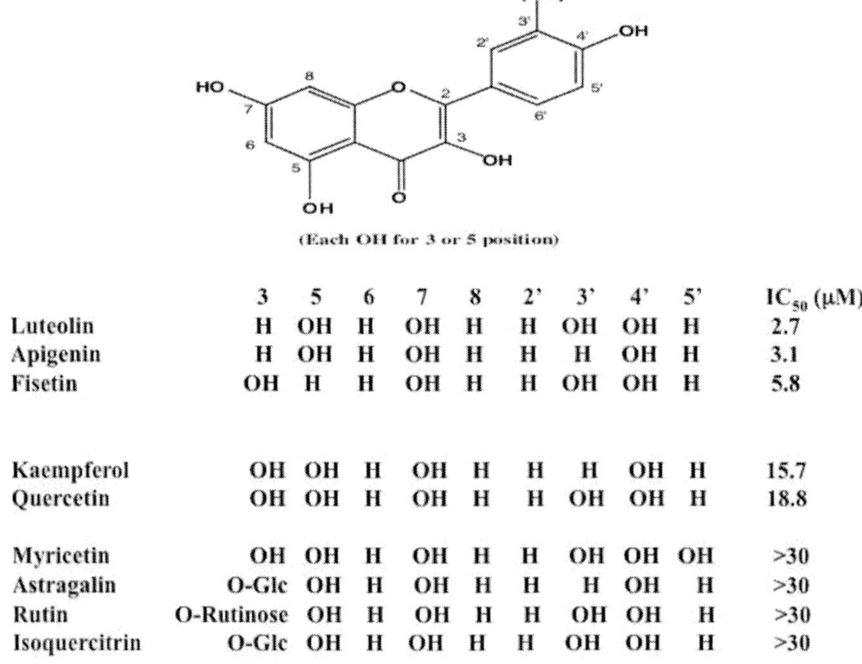

	3	5	6	7	8	2'	3'	4'	5'	IC_{50} (μM)
Luteolin	H	OH	H	OH	H	H	OH	OH	H	2.7
Apigenin	H	OH	H	OH	H	H	H	OH	H	3.1
Fisetin	OH	H	H	OH	H	H	OH	OH	H	5.8
Kaempferol	OH	OH	H	OH	H	H	H	OH	H	15.7
Quercetin	OH	OH	H	OH	H	H	OH	OH	H	18.8
Myricetin	OH	OH	H	OH	H	H	OH	OH	OH	>30
Astragalin	O-Glc	OH	H	OH	H	H	H	OH	H	>30
Rutin	O-Rutinose	OH	H	OH	H	H	OH	OH	H	>30
Isoquercitrin	O-Glc	OH	H	OH	H	H	OH	OH	H	>30

Figure 1. Basic structure of flavonoids for inhibitory activity on IL-4 expression by basophils. Luteolin, fisetin and apigenin are the strongest inhibitors.

The mechanisms through which flavonoids inhibit IL-4 and CD40 ligand expression by basophils are not yet fully understood, but in studies conducted by us, flavonoids suppressed activation of transcriptional factors such as AP-1 and the nuclear factor of activated T cells (NFAT)-1 by KU812 cells [20, 24]. Several findings have been reported regarding the action of flavonoids on basophils or mast cells. Kimata et al showed that luteolin and quercetin inhibited Ca^{2+} influx, PKC activity, and the activation of p44/42 MAPK and JNK activation but not of p38 MAPK by human cultured mast cells in response to cross-linkage of FcεRI [17], whereas quercetin reportedly attenuates p38 MAPK and NF-kB in HMC-1, a human mast cell line [25]. Shichijo et al further demonstrated that flavonoids inhibited Syk activation in human cultured mast cells as well in response to cross-linkage of FcεRI [26]. However, the relationship between the inhibition of Syk activation and degranulation remains unclear, since myricetin proved to be a strong inhibitor of Syk activation but a weak suppressor of degranulation. In another study, flavones including apigenin were found to reduce the FcεRI α and γ chains, perhaps through down-regulating P44/42 MAPK phosphorylation in KU812 cells [27]. The reasons for the discrepancy among these previously reported findings regarding flavonoids' suppressive activity on the activation of MAPK family members are not known at present, but it may be due to differences in cells, stimulation, or flavonoid doses, or it may be due to the presence of several acting points of flavonoids. Further studies are thus required to clarify these points. Finally, kaempferol was recently reported to inhibit IL-4-induced STAT6 activation by specifically targeting JAK3 in hemapoietic cell lines, which thus constitutes another anti-allergic activity of flavonoids [28].

II. Clinical Effects of Flavonoids Identified with in Vivo Animal Models

Based on the anti-allergic characteristics of flavonoids observed *in vitro*, NC/Nga mouse was used to test whether intake of flavonoids might be effective for the amelioration of allergic symptoms or for prevention of the onset of allergic diseases. NC/Nga mice spontaneously develop severe eczema, scratching behaviour and serum IgE elevation with aging under nonspecific pathogen-free conditions [29]. To determine the preventive effect of flavonoids, the mice were orally given astragalin, kaempferol 3'glucoside (1.5 mg/kg) or a control diet. Development of dermatitis with aging was observed in the control group and the severity of dermatitis was evaluated with a score. Oral intake of astragalin markedly prevented the appearance of the skin symptoms, scratching behaviour and serum IgE elevation [30]. Moreover, when astragalin was administered to NC/Nga mice after the appearance of dermatitis, it significantly diminished its severity [31]. Similarly, it was subsequently demonstrated with this mouse model that administration of an extract from petals of *Impatiens balsamina* L. including flavonoids such as kaempferol 3-rutinoside and 2-hydroxy-1,4-naphthoquinone as active gradients, suppressed scratching behaviour and dermatitis [32]. Moreover, in an asthmatic model mouse sensitized with ovalbumin, it was demonstrated that oral administration of luteolin, even as little as 0.1 mg/kg, led to a significant suppression of bronchial hyperreactivity and bronchoconstriction [33]. It was recently reported that nobiletin, a polymethoxyflavonoid, when given intraperitoneally to OVA-sensitized rat at a dose of 1.5 or 5 mg/kg, reduced OVA-induced increases in eosinophils and eotaxin expression [34]. Flavonoids such as quercetin, isoquercitrin, rutin, 3-O-methylquerctin 5,7,3',4'-O-tetraacetate, narirutin (naringenin-7-O-β-D-rutiose) and apigenin have subsequently been shown to suppress allergic responses in several types of allergic animals [35-39].

III. Flavonoids Ameliorated Ocular Symptoms of Japanese Cedar Pollinosis

The findings regarding *in vitro* or *in vivo* anti-allergic properties of flavonoids are summarized. Flavonoids, in particular luteolin, apigenin and fisetin, have strong anti-allergic effects such as inhibition of histamine release, of IL-4 and IL-13 synthesis and of CD40 ligand expression by basophils. The inhibitory activity of flavonoids against IL-4, IL-13 and CD40 ligand expression may be mediated through their inhibition of transcriptional factors such as AP-1 and NFAT-1. In allergic animal models, administration of flavonoids has been shown to be effective as treatment or prevention. One epidemiological report of a 30-year longitudinal study mentioned that the incidence of asthma is lower in populations with a higher intake of flavonoids and that the relative risk was 0.65 [40]. This evidence strongly supports the notion that an appropriate intake of flavonoids may constitute a complementary and alternative medicine and/or a preventive strategy for allergic diseases [22, 41, 42].

Enzymatically modified isoquercitrin (EMIQ) is a quercetin glycoside that consists of isoquercitrin and its maltooligosaccharides, and is manufactured from rutin through an enzymatic modification (Figure 2) [43], which markedly enhances the absorption rate through intestine. This flavonoid is recognized as a food additive and is used in Japan for various commercially available food products such as beverages as an anti-oxidant which is highly soluble in water.

Tests were performed in 2007 and 2008 to determine whether intake of EMIQ was effective for allergic diseases. In a parallel-group, double-blind, placebo-controlled study, volunteers with Japanese cedar pollinosis took two capsules of 100mg EMIQ or a placebo for 8 weeks during the pollen season. The study in 2007 began after the pollen dispersed and thus aimed at examining the therapeutic effect of EMIQ. Although total symptom (nasal and ocular symptoms)+medication and total nasal+medication scores were not significantly reduced during the entire period, the total ocular+medication score for the EMIQ group was significantly lower (Figure 2) [44]. In 2008, the preventive effect of EMIQ on symptoms of pollinosis was also examined and the study began 3 weeks before the first day of pollen dispersion. The results of the study revealed that intake of EMIQ also led to an amelioration of ocular but not nasal symptoms caused by pollinosis.

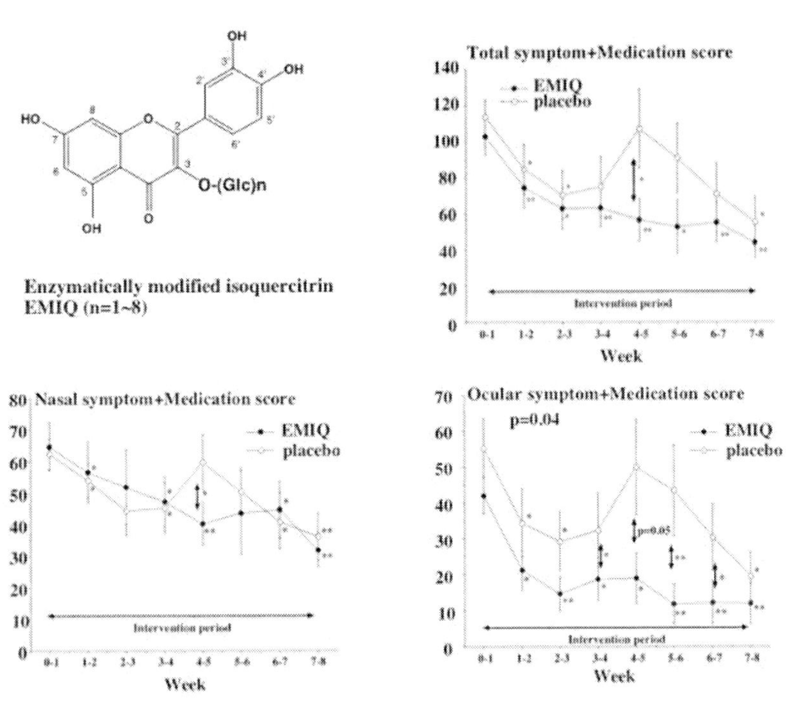

Figure 2. Structure of enzymatically modified isoquercitrin (EMIQ) and the effect of EMIQ on symptoms of Japanese cedar pollinosis. The changes of total symptom+medication, nasal symptom+medication and ocular symptom+medication scores in the EMIQ (closed circle) and the placebo (open circle) were shown. Each value is the mean +/- SE. *p<0.05, **p<0.01compared with the first week in each group. Arrow with *p<0.05, **p<0.01 compared both groups in the same week.

Conclusion

Allergy, a worldwide common disease, is the subject of growing concern because of its increasing rate of prevalence [45]. It has been suggested that dietary changes may contribute to this increase [46, 47]. Flavonoids such as luteolin, apigenin, fisetin, quercetin and kaempferol) have anti-allergic effects that inhibit histamine, IL-4, IL-13 and CD40 ligand expression by basophils. The database of the flavonoid content of major vegetables, fruits and drinks [48] is expected to contribute to an appropriate intake of flavonoids from foods and drinks to ameliorate allergic symptoms and prevent the onset of allergic diseases.

Acknowledgments

The authors gratefully acknowledge the collaboration and assistance of Naoyoshi Maezaki, Tetsuaki Tanaka, Hisashi Matsuda, Masayuki Yoshikawa, Masahiro Takigawa, Katuyori Kouda, Hirokazu Takeuchi, Kana Ioku, Mituo Kouda, Mayumi Kotani, Akihito Fujita, Motonobu Matsumoto, Akira Takeuchi, Toshio Tabei, Masamitsu Moriwaki and Yukio Suzuki.

References

[1] Falcone, F. H., Haas, H. & Gibbs, B. F. (2000). The human basophils: a new appreciation of its role in immune responses. *Blood*, *96*, 4028-4038.

[2] Gibbs, B. F. (2005). Human basophils as effectors and immunomodulators of allergic inflammation and innate immunity. *Clin Exp Med*, *5*, 43-49.

[3] Yanagihara, Y. (2003). Regulatory mechanisms of human IgE synthesis. *Allergology International*, *52*, 1-12.

[4] Gibbs, B. F., Vollrath, I. B., Albrecht, C., Amon, U. & Wolff, H. H. (1998). Inhibition of interleukin-4 and interleukin-13 release from immunologically activated human basophils due to the actions of anti-allergic drugs. *Naunyn-Schmiederberg's Arch Pharmacol*, *357*, 573-578.

[5] Tanaka, T., Kouda, K., Kotani, M., Takeuchi, A., Tabei, T., Masamoto, Y., Nakamura, H., Takigawa, M., Suemura, M., Takeuchi, H. & Kouda, M. (2001). Vegetarian diet ameliorates symptoms of atopic dermatitis through reduction of the number of peripheral eosinophils and of PGE2 synthesis by monocytes. *J Physiol Anthropol Appl Human Sci*, *20*, 353-361.

[6] Kouda, K., Tanaka, T., Kouda, M., Takeuchi, H., Takeuchi, A., Nakamura, H. & Takigawa, M. (2000). Low-energy diet in atopic dermatitis patients: clinical findings and DNA damage. *J Physiol Anthropol Appl Human Sci*, *19*, 225-228.

[7] Hollman, P. C. & Katan, M. B. (1999). Health effects and bioavailability of dietary flavonols. *Free Rad Res*, *31 Suppl*, S75-80.

[8] Harborne, J. B. & Williams, C. A. (2000). Advances in flavonoid research since 1992. *Phytochemistry*, *55*, 481-504.

[9] Middleton, E. J., Kandaswami, C. & Theoharides, T. C. (2000). The effects of plant flavonoids on mammalian cells: implications for inflammation, heart disease, and cancer. *Pharmacol Rev*, *52*, 673-751.

[10] Williams, C. A. & Grayer, R. J. (2004). Anthocyanins and other flavonoids. *Nat Prod Rep*, *21*, 539-573.

[11] Fewtrell, C. M. & Gomperts, B. D. (1997). Effect of flavone inhibitors on transport ATPases on histamine secretion from rat mast cells. *Nature*, *265*, 635-636.

[12] Middleton, E. J., Drzewiecki, G. & Krishnarao, D. (1981). Quercetin: an inhibitor of antigen-induced human basophil histamine release. *J Immunol*, *127*, 546-550.

[13] Middleton, E. J. & Kandaswami, C. (1992). Effects of flavonoids on immune and inflammatory cell functions. *Biochem Pharmacol*, *43*, 1167-1179.

[14] Cheong, H., Ryu, S. Y., Oak, M. H., Cheon, S. H., Yoo, G. S. & Kim, K. M. (1998). Studies of structure activity relationship of flavonoids for the anti-allergic actions. *Arch Pharm Res*, *21*, 478-480.

[15] Lee, T. P., Matteliano, M. L. & Middleton, E. J. (1982). Effect of quercetin on human polymorphonuclear leukocyte lysosomal enzyme release and phospholipid metabolism. *Life Sci*, *31*, 2765-2774.

[16] Yoshimoto, T., Furukawa, M., Yamamoto, S., Horie, T. & Watanabe-Kohno, S. (1983). Flavonoids: potent inhibitors of arachidonate 5-lipoxygenase. *Biochem Biophys Res Commun*, *116*, 612-618.

[17] Kimata, M., Shichijo, M., Miura, T., Serizawa, I., Inagaki, N. & Nagai, H. (2000). Effects of luteolin, quercetin and baicalein on immunoglobulin E-mediated mediator release from human cultured mast cells. *Clin Exp Allergy*, *30*, 501-508.

[18] Kimata, M., Inagaki, N. & Nagai, H. (2000). Effects of luteolin and other flavonoids on IgE-mediated allergic reactions. *Plant Med*, *66*, 25-29.

[19] Gauchat, J. F., Henchoz, S., Mazzei, G., Aubry, J. P., Brunner, T., Blasey, H., Life, P., Talabot, D., Flores-Romo, L., Thompson J., Kishi, K.,

Butterfield, J., Dahinden, C. & Bonnefoy, J. Y. (1993). Induction of human IgE synthesis in B cells by mast cells and basophils. *Nature*, *365*, 340-343.

[20] Higa, S., Hirano, T., Kotani, M., Matsumoto, M., Fujita, A., Suemura, M., Kawase, I. & Tanaka, T. (2003). Fisetin, a flavonol, inhibits TH2-type cytokine production by activated human basophils. *J Allergy Clin Immunol*, *111*, 1299-1306.

[21] Hirano, T., Higa, S., Arimitsu, J., Naka, T., Shima, Y., Ohshima, S., Fujimoto, M., Yamadori, T., Kawase, I. & Tanaka, T. (2004). Flavonoids such as luteolin, fisetin and apigenin are inhibitors of interleukin-4 and interleukin-13 production by activated human basophils. *Int Arch Allergy Immunol*, *134*, 135-140.

[22] Kawai, M., Hirano, T., Higa, S., Arimitsu, J., Maruta, M., Kuwahara, Y., Ohkawara, T., Hagihara, K., Yamadori, T., Shima, Y., Ogata, A., Kawase, I. & Tanaka, T. (2007). Flavonoids and related compounds as anti-allergic substances. *Allergology International*, *56*, 113-123.

[23] Hirano, T., Arimitsu, J., Higa, S., Naka, T., Ogata, A., Shima, Y., Fujimoto, M., Yamadori, T., Ohkawara, T., Kuwahara, Y., Kawai, M., Kawase, I. & Tanaka, T. (2006). Luteolin, a flavonoid, inhibits CD40 ligand expression by activated human basophils. *Int Arch Allergy Immunol*, *140*, 150-156.

[24] Hirano, T., Higa, S., Arimitsu, J., Naka, T., Ogata, A., Shima, Y., Fujimoto, M., Yamadori, T., Ohkawara, T., Kuwahara, Y., Kawai, M., Matsuda, H., Yoshikawa, M., Maezaki, N., Tanaka, T., Kawase, I. & Tanaka, T. (2006). Luteolin, a flavonoid, inhibits AP-1 activation by basophils. *Biochem Biophys Res Commun*, 340, 1-7.

[25] Min, Y. D., Choi, C. H., Bark, H., Son, H. Y., Park, H. H., Lee, S., Park, J. W., Shin, H. I. & Kim, S. H. (2007). Quercetin inhibits expression of inflammatory cytokines through attenuation of NF-kappaB and p38 MAPK in HMC-1 human mast cell line. *Inflamm Res*, *56*, 210-215.

[26] Shichijo, M., Yamamoto, N., Tsujishita, H., Kimata, M., Nagai, H. & Kokubo, T. (2003). Inhibition of syk activity and degranulation of human mast cells by flavonoids. *Biol Pharm Bull*, *26*, 1685-1690.

[27] Yano, S., Tachibana, H. & Yamada, K. (2005). Flavones suppress the expression of the high-affinity IgE receptor Fc epsilon RI in human basophilic KU812 cells. *J Agric Food Chem*, *53*, 1812-1817.

[28] Cortes, J. R., Perez-G, M., Rivas, M. D. & Zamorano, J. (2007). Kaempferol inhibits IL-4-induced STAT6 activation by specifically targeting JAK3. *J Immunol*, *179*, 3881-3887.

[29] Matsuda, H., Watanabe, N., Geba, G. P., Sperl, J., Tsudzuki, M., Hiroi, J., Matsumoto, M., Ushio, H., Saito, S., Askenase, P. W. & Ra, C. (1997).

Development of atopic dermatitis-like skin lesion with IgE hyperproduction in NC/Nga mice. *Int Immunol*, *9*, 461-466.

[30] Kotani, M., Matsumoto, M., Fujita, A., Higa, S., Wang, W., Suemura, M., Kishimoto, T. & Tanaka, T. (2000). Persimmon leaf extract and astragalin inhibit development of dermatitis and IgE elevation in NC/Nga mice. *J Allergy Clin Immunol*, *106*, 159-166.

[31] Matsumoto, M., Kotani, M., Fujita, A., Higa, S., Kishimoto, T., Suemura, M. & Tanaka, T. (2002). Oral administration of persimmon leaf extract ameliorates skin symptoms and transepidermal water loss in atopic dermatitis model mice, NC/Nga. *Brit J Dermatol*, *146*, 221-227.

[32] Oku, H. & Ishiguro, K. (2001). Antipruritic and antidermatitic effects of extract and compounds of *Impatiens balsamina* L. in atopic dermatitis model NC mice. *Phytother Res*, *15*, 506-510.

[33] Das, M., Ram, A. & Ghosh, B. (2003). Luteolin alleviates bronchoconstriction and airway hyperreactivity in ovalbumin sensitized mice. *Inflamm Res*, *52*, 101-106.

[34] Wu, Y. Q., Zhou, C. H., Tao, J. & Li, S. N. (2006). Antagonistic effects of nobiletin, a polymethoxyflavonoid, on eosinophilic airway inflammation of asthmatic rats and relevant mechanisms. *Life Sci*, *78*, 2689-2696.

[35] Rogerio, A. P., Kanashiro, A., Fontanari, C., da Silva, E. V., Lucisano-Valim, Y. M., Soares, E. G. & Faccioli, L. H. (2007). Anti-inflammatory activity of quercetin and isoquercitrin in experimental murine allergic asthma. *Inflamm Res*, *56*, 402-408.

[36] Jung, C. H., Lee, J. Y., Cho, C. H. & Kim, C. J. (2007). Anti-asthmatic action of quercetin and rutin in conscious guinea-pigs challenged with aerosolized ovalbumin. *Arch Pharm Res*, *30*, 1599-1607.

[37] Jiang, J. S., Chien, H. C., Chen, C. M., Lin, C. N. & Ko, W. C. (2007). Potent suppressive effects of 3-0-methylquercetin 5,7,3',4'-O-tetraacetate on ovalbumin-induced airway hyperresponsiveness. *Planta Med*, *73*, 1156-1162.

[38] Funaguchi, N., Ohno, Y., La, B. L., Asai, T., Yuhgetsu, H., Sawada, M., Takemura, G., Minatoguchi, S., Fujiwara, T. & Fujiwara, H. (2007). Narirutin inhibits airway inflammation in an allergic mouse model. *Clin Exp Pharmacol Physiol*, *34*, 766-770.

[39] Yano, S., Umeda, D., Yamashita, S., Ninomiya, Y., Sumida, M., Fujimura, Y., Yamada, K. & Tachibana, H. (2007). Dietary flavones suppress IgE and Th2 cytokines in OVA-immunized BALB/c mice. *Eur J Nutr*, *46*, 257-263.

[40] Knekt, P., Kumpulainen, J., Jarvinen, R., Rissanen, H., Heliovaara, M., Reunanen, A., Hakulinen, T. & Aromaa A. (2002). Flavonoid intake and

risk of chronic diseases. *Am J Clin Nutr*, *76*, 560-568.

[41] Tanaka, T., Higa, S., Hirano, T., Kotani, M., Matsumoto, M., Fujita, A. & Kawase, I. (2003). Flavonoids as potential anti-allergic substances. *Curr Med Chem-Anti-Inflammatory & Anti-Allergy Agents*, *2*, 57-65.

[42] Tanaka, T., Higa, S., Hirano, T., Arimitsu, J., Naka, T., Shima, Y., Ohshima, S., Fujimoto, M., Yamadori, T. & Kawase, I. (2004). Is an appropriate intake of flavonoids a prophylactic means or complementary and alternative medicine for allergic diseases? *Recent Res Devel Allergy & Clin Immunol*, *5*, 1-14.

[43] Toyoshi, T., Emura, K. & Moriwaki, M. (2005). Antihypertensive effect of enzymatically modified isoquercitrin in spontaneously hypertensive rats. *Foods & Food Ingredients J Japan*, *210*, 778-783.

[44] Kawai, M., Hirano, T., Arimitsu, J., Higa, S., Kuwahara, Y., Hagihara, K., Shima, Y., Narazaki, M., Ogata, A., Koyanagi, M., Kai, T., Shimizu, R., Moriwaki, M., Suzuki, Y., Ogino, S., Kawase, I. & Tanaka, T. (2009). Enzymatically modified isoquercitrin, a flavonoid, on symptoms of Japanese cedar pollinosis: a randomized double-blind placebo-controlled trial. *Int Arch Allergy immunol*, in press.

[45] Holgate, S. T. (1999). The epidemic of allergy and asthma. *Nature*, *402*, B2-4.

[46] McKeever, T. M. & Britton, J. (2004). Diet and asthma. *Am J Respir Crit Care Med*, *170*, 725-729.

[47] Devereux, G. (2006). The increase in the prevalence of asthma and allergy: food for thought. *Nat Rev Immunol*, *6*, 869-874.

[48] US Department of Agriculture. USDA database for the flavonoid content of selective foods. [Cited 2003 March]. Available from http://www.nal.usda.gov/fnic/foodcomp/

In: Basophil Granulocytes
Editors: Paul K. Vellis

ISBN: 978-1-60741-797-2

Chapter 5

BASOPHIL IN TROPICAL INFECTIONS

Viroj Wiwanitkit
Wiwanitkit House, Bangkhae, Bangkok Thailand 10160

ABSTRACT

Basophils are a minute group of granulocytes. Among types of white blood cells, they are the least in number. Knowledge on the basophil is limited. In this article, the author will briefly detail and discuss basophils in relation to tropical infections. The scenarios of malaria, dengue, tuberculosis as well as leprosy will be presented.

INTRODUCTION

Basophils are a minute group of granulocytes. Their level is about 0–1 % of all white blood cells—the least in number among types of white blood cells. Knowledge on basophils is limited. Most of the reseach done on basophils involves hypersensitivity and allergy. However, basophils also play important roles in other disoders. Infection is mentioned as having a possible clinical correlation to basophils. Knowledge about this area would be of interest. In this article, the author will briefly detail and discuss basophils in relation to tropical infections. The scenarios of malaria, dengue, tuberculosis as well as leprosy will be presented.

Basophil in Malaria

Malaria is an important tropical mosquito-borne infectious disease. It is a parasitic infection. The protozoa in the species of Plasmodium is the pathogenic organism. This organism directly hits the red blood cells and causes the pathology. Malaria is still prevalent in many tropical countries around the world. Thousands of malarial research studies are in the medical literature. However, there are only a few reports on basophils in malaria. Kojima et al. summarized the role of interleukin (IL) 18 in severe falciparum malaria. Basically, IL-18 is a strong proinflammatory cytokine that induces interferon-gamma (IFN-gamma) production from Th1 cells, NK cells and activated macrophages, particularly in the existence of IL-12, and this IL is capable of inducing IL-4 and IL-13 production in T cells, NK cells, mast cells and basophils, and administration of IL-18 in conjunction with an allergen increases serum IgE levels [1]. Tsutsui et al. said that IL-18 also had the capacity to induce allergic responses by specific induction of IL-4 production by T helper cells and to activate mast cells and basophils to release atopic effector molecules such as histamine [2]. Tsutsui et al. said that this related to malarial hepatitis [2]. Nyakeriga et al. also noted that human basophils might contribute to the polarization of T-helper type 2 in the (Th2) responses in malaria hosts by IgE-induced IL-4 generation [3]. Poorafshar et al. showed that basophil-like cells were very potent producers of IL-4 and that IL-4 produced by these cells could be of main importance for the initiation of a Th2 response [4]. In addition, MacDonald proposed that malarial *Plasmodium falciparum* translationally controlled tumor protein, presented in human plasma during a malarial illness, might affect host immune responses in vivo [5]. Indeed, this protein is proved to be a homolog of the mammalian histamine-releasing factor (HRF), which causes histamine release from human basophils and IL-8 secretion from eosinophils [5]. Finally, Kamis and Ibrahim studied effects of testosterone on blood leukocytes in plasmodium berghei-infected mice [6]. They reported that the number of all individual types of leukocytes except basophils in vehicle-treated gonadectomized mice was increased [6].

Basophil in Dengue

Similar to malaria, dengue is an important tropical mosquito-borne infection. The public health impact of this arboviral infection is as high as that of malaria. The acute fever accompanied with deadly bleeding episode is the hallmark for this viral infection. A high rate of fatality can be seen if there is no proper medical management. At present, dengue infection is still prevalent in many tropical countries around the world. Thousands of dengue researches are in the medical literatures. However, there are only a few reports on basophils in dengue infection. King et al. suggested a role for mast cells in the initiation of chemokine-dependent host responses to dengue virus infection [7]. A selective mast cell response to dengue was observed in the research of King et al. [7]. Release of vasoactive cytokines by antibody-enhanced dengue virus infection of a human mast cell/basophil line was also observed [8]. Liu et al. proposed interesting information on automated hematological appearance in dengue infection [9]. Liu et al. found that Technicon H*1 derived basophil and large unstained cell counts and manual atypical lymphocyte counts rose at the same time as the decrease in platelets and decreased when the platelets recovered [9]. Indeed, dramatic caspase-dependent apoptosis in antibody-enhanced dengue virus infection of human mast cells was observed by Brown et al. [10]. Brown et al. found that antibody-enhanced dengue virus infection of the FcR-bearing mast cell/basophil KU812 cell line could bring a massive induction of apoptosis [10]. However, the involvement of basophil and dengue hemorrhagic fever, the serious dengue infection via endothelial involvement is not noted [11].

BASOPHIL IN TUBERCULOSIS

Tuberculosis is one of the problematic chronic respiration diseases. This infection is caused by *Mycobacterium tuberculosis*. This is an old infection. However, owing to the emergence of HIV infection, tuberculosis becomes a focused remerging infectious disease. At present, tuberculosis is still problematic in many countries around the world. Thousands of tuberculosis research studies are in the medical literature. However, there are only a few reports on basophils in tuberculosis. Tuberculosis is the cause of basophilic pleural effusion [12]. The value of the basophil test in differential diagnosis of adenopathies of tuberculous and nonspecific etiology in children was also approved [13]. The patients with cavitary lesions showed significantly higher levels of anti-TBGL IgG, anti-TBGL IgA , white blood cells , neutrophils, basophils, natural killer cells, CRP, KL-6 (sialylated carbohydrate antigen KL-6) , IgA , and sCD40L [14].

The basophil in the hypersensitivity in tuberculosis, namely, a Jones-Mote reaction is also noted. Basically, Jones-Mote reaction is mentioned as a delayed, erythematous, and mildly indurated cutaneous reaction originally described in humans sensitized by skin injection of heterologous proteins [15]. This can be seen in the tuberculin skin test. Nakamura et al. proposed that macrophages activated by complete Freund's adjuvant might have an important role in regulating basophil infiltration in the effector phase of the delayed hypersensitivity reaction [16]. Askenase and Atwood suggested that the occurrence of basophils at delayed reactions was owing to complex regulation and that basophil accumulations are an aspect of delayed hypersensitivity, rather than an indication of a distinctive and separate response [15].

Basophil in Leprosy

Leprosy is another important mycobacterium infection. This disease is still prevalent in the setting with poor sanitation. Disability due to leprosy is very high and can be a public health problem. Thousands of leprosy research studies are in the medical literature. However, there are only a few reports on basophils in leprosy.

Samuel et al. studied microscopic findings of delayed reactions elicited by the skin test reagent Leprosin A derived from *Mycobacterium leprae* [16]. They noted that the most striking findings were the absence of the expected basophils and an infiltration of eosinophils which proceeded to degranulate [16].

References

[1] Kojima, S; Nagamine, Y; Hayano, M; Looareesuwan, S; Nakanishi, K. A potential role of interleukin 18 in severe falciparum malaria. *Acta Trop.*, 2004 Feb, 89(3), 279-84.

[2] Tsutsui, H; Adachi, K; Seki, E; Nakanishi, K. Cytokine-induced inflammatory liver injuries. *Curr Mol Med.*, 2003 Sep, 3(6), 545-59.

[3] Nyakeriga, MA; Troye-Blomberg, M; Bereczky, S; Perlmann, H; Perlmann, P; ElGhazali, G. Immunoglobulin E (IgE) containing complexes induce IL-4 production in human basophils: effect on Th1-Th2 balance in malaria. *Acta Trop.*, 2003 Apr, 86(1), 55-62.

[4] Poorafshar, M; Helmby, H; Troye-Blomberg, M; Hellman, L. MMCP-8, the first lineage-specific differentiation marker for mouse basophils. Elevated numbers of potent IL-4-producing and MMCP-8-positive cells in spleens of malaria-infected mice. *Eur J Immunol.*, 2000 Sep, 30(9), 2660-8.

[5] MacDonald, SM; Bhisutthibhan, J; Shapiro, TA; Rogerson, SJ; Taylor, TE; Tembo, M; Langdon, JM; Meshnick, SR. Immune mimicry in malaria: Plasmodium falciparum secretes a functional histamine-releasing factor homolog in vitro and in vivo. *Proc Natl Acad Sci*, USA., 2001 Sep 11, 98(19), 10829-32.

[6] Kamis, AB; Ibrahim, JB. Effects of testosterone on blood leukocytes in plasmodium berghei-infected mice. *Parasitol Res*., 1989, 75(8), 611-3.

[7] King, CA; Anderson, R; Marshall, JS. Dengue virus selectively induces human mast cell chemokine production. *J Virol*., 2002 Aug, 76(16), 8408-19.

[8] King, CA; Marshall, JS; Alshurafa, H; Anderson, R. Release of vasoactive cytokines by antibody-enhanced dengue virus infection of a human mast cell/basophil line. *J Virol*., 2000 Aug, 74(15), 7146-50.

[9] Liu, TC; Chan, YC; Han, P. Lymphocyte changes in secondary dengue fever: use of the Technicon H*1 to monitor progress of infection. *Southeast Asian J Trop Med Public Health*., 1991 Sep, 22(3), 332-6.

[10] Brown, MG; Huang, YY; Marshall, JS; King, CA; Hoskin, DW; Anderson, R. Dramatic caspase-dependent apoptosis in antibody-enhanced dengue virus infection of human mast cells. *J Leukoc Biol*., 2008 Sep 22. [Epub ahead of print]

[11] Butthep, P; Bunyaratvej, A; Bhamarapravati, N. Dengue virus and endothelial cell: a related phenomenon to thrombocytopenia and granulocytopenia in dengue hemorrhagic fever. *Southeast Asian J Trop Med Public Health*., 1993, 24 Suppl 1, 246-9.

[12] Okimoto, N; Kurihara, T; Honda, Y; Ohba, H. Cause of basophilic pleural effusion. *South Med J*., 2003 Jul, 96(7), 726-7.

[13] Borris, VM; Shubskaia, OI. Value of the basophil test in differential diagnosis of adenopathies of tuberculous and nonspecific etiology in children. *Probl Tuberk*., 1978 Jan, (1), 58-61.

[14] Misuzawa, M; Kawamura, M; Takamori, M; Kashiyama, T; Fujita, A; Usuzawa, M; Saitoh, H; Ashino, Y; Yano, I; Hattori, T. Increased synthesis of anti-tuberculous glycolipid immunoglobulin G (IgG) and IgA with cavity formation in patients with pulmonary tuberculosis. *Clin Vaccine Immunol*., 2008 Mar, 15(3), 544-8.

[15] Askenase, PW; Atwood, JE. Basophils in tuberculin and "Jones-Mote" delayed reactions of humans. *J Clin Invest*., 1976 Nov, 58(5), 1145-54.

[16] Nakamura, S; Sanui, H; Nomoto, K. Relationship between the tuberculin-type and Jones-Mote-type hypersensitivities: suppression of basophil infiltration by mycobacterial adjuvant. *Immunology*., 1986 Jul, 58(3),

397-403.

[17] Samuel, NM; Stanford, JL; Desikan, KV. Microscopic findings of delayed reactions elicited by the skin test reagent Leprosin A derived from M. leprae. *Int J Lepr Other Mycobact Dis*., 1985 Sep, 53(3), 395-403.

In: Basophil Granulocytes
Editor: Paul K. Vellis

ISBN: 978-1-60741-797-2

Chapter 6

RESEARCH ON BASOPHIL IN THAILAND

Viroj Wiwanitkit
Wiwanitkit House, Bangkhae, Bangkok Thailand 10160

ABSTRACT

Until now, the basophil had been among the least studied types of white blood cells. Basically, medical scientists know that the main role of the basophil relates to an anaphylaxis reaction. However, there are many other facets of basophil roles. In this chapter, the author will briefly summarize important reports on the basophil in Thailand.

INTRODUCTION

The basophil is a type of white blood cell containing large basophilic granules. Until now, the basophil had been studied less than other types of white blood cells. Basically, medical scientists know that the main role of the basophil relates to an anaphylaxis reaction. However, there are many other facets of basophil roles, and those roles are of interest. The author will briefly summarize important reports done in Thailand on the basophil.

Basophil in Allergy

As already mentioned, the basophil is known for its role in the anaphylaxis process, the hypersensitivity type 1 reaction. Research on the basophil relating to allergy is common. In Thailand, there are also some reports on basophil in allergy. Innajak and Srimarut studied asthmatic patients [1]. They studied 9 healthy subjects compared to 8 asthmatic cases [1]. In their work, the histamine level of the normal group was equal to 17.46 +/- 6.71 ng/ml, whereas the level of the asthmatic group was equal to 59.98 +/- 16.14 ng/ml [1]. There is a statistically significant difference in the histamine level between both groups [1]. Innajak and Srimarut also found that there was a significant correlation between basophil count and histamine level [1]. Sitthidilokrat performed a study on cell type of pediatric patients with allergic rhinitis [2]. Sitthidilokrat studied 80 cases and found that the level of basophil was about 1 + and about 74% [2]. Co-presentation with eosinophil was also noted [2].

Jirapongsananuruk and Vichyanond performed a similar study and noted that the sensitivity for nasal eosinophil scores or nasal basophilic metachromatic cell scores more than 0.5 in the diagnosis of allergic rhinitis was 91.7% with a specificity of 100%, positive predictive value of 100% and a negative predictive value of 91.1% [3]. Of interest, Janwitayanujit also said that anaphylactoid were not mediated by antigen-antibody but result from substances acting directly on mast cells and basophils [4]. Janwitayanujit proposed that anaphylaxis could spontaneously occur with no external allergen [4]. Concerning allergic dermatological manifestation, Kulthanan et al. studied the prevalence of autologous serum skin test (ASST)—the best in vivo clinical test for the detection of basophil histamine-releasing activity in vitro—positive cases in Thai patients with chronic idiopathic urticaria (CIU) and identified factors related to the positivity of ASST and looked for the clinical implications of ASST in CIU [5]. Kulthanan et al. concluded that there was no significant difference between patients with positive ASST and negative ASST as to the severity of the disease (wheal numbers, wheal size, itching scores and the extent of body involvement) as well as the duration of the disease [5]. Finally, Wiwanitkit used a new advanced nanomedical technique to study the basophil [6]. Wiwanitkit detected the transmembrane oxidation flux in the basophil by using a simulation test to determine the oxidation flux change based on nanomedicine technique [6]. According to this work, no change of flux could be determined [6]. Wiwanitkit said that the finding in this research could support the finding that the oxidation flux change was not an important part in the pathogenesis of basophil-related hypersensitivity [6].

Basophil in Cancer

The role of the basophil in cancer is the present focus in oncology. There are many research studies on this area around the world, including in Thailand. Palungwachira et al. reported on a patient with solitary mastocytoma with a solitary red infiltrated plaque on the dorsum of the right foot for 2 months [7]. In this case, electron microscopy revealed Charcot-Leyden crystals (CLCs) located in phagosomes of activated macrophages as well as in the stromal tissue; a close association between CLCs formation and damaged eosinophils could be seen [7]. Palungwachira et al. showed the evidence that the formation of CLCs in a mastocytoma correlated to the individual and related to the biology of mast cells, basophils, eosinophils and macrophages and noted that phagosomes might have acted as the localization of CLCs formation [7]. Tirakunwicha and Tulvatana studied the basophil in another cancer, adenoid cystic carcinoma [8]. Tirakunwicha and Tulvatana also reported that the primitive cells (basaloid pattern) disappeared after orbital irradiation for adenoid cystic carcinoma of the lacrimal gland due to the replacement with fibrous tissue [8].

Basophil in Infection

The role of basophil in infection has a limited focus. However, there are some interesting reports on the basophil in infection. In Thailand, there was an interesting paper describing the basophil in dengue infection, an important tropical arboviral infection [9]. Butthep et al. proposed that monocytes, basophils and eosinophils had no interaction with dengue-infected endothelial cells in the pathogenesis [9]. Butthep et al. said that the increased binding of neutrophil and platelet to endothelial cell was the clue explaining neutropenia and thrombocytopenia in patients with dengue hemorrhagic fever [9].

Basophil in Intoxication

The basophil in intoxication disorder is not widely mentioned. However, there are some interesting reports on the basophil in intoxication disorder. Salakij et al. performed an interesting research to determine basic hematologic values and evaluate light microscopic, cytochemical, and electron microscopic characteristics of king cobra blood cells [10]. Salakij et al. said that basophil granules could be stained with PAS, SBB, ANAE, and beta-glu [10]. Salakij et al. presented some theories on the effect of toxin from the king cobra on human beings [10]. Wiwanitkit studied the basophil in relation to exposure to benzene [11]. Wiwanitit et al. studied 25 at-risk subjects [11]. Wiwanitkit et al. reported that there was no siginificant correlation between the standard biomarker for benzene exposure, urine phenol level and the basophil count [11].

What Should the Next Trend of Research in Thailand Be?

It can be seen that there are only a few research studies on the basophil in Thailand. There is a need for continuous study on the basophil. The author thinks that the trend of the basophil research in Thailand should be studies of tropical diseases. There are many interesting conditions that could be studied. An example is allergy due to tropical plants which are common in Thailand.

References

[1] Innajak, K; Srimarut, N. Histamine, basophil and eosinophil level in blood of astmatic patients. *Siriraj Hosp Gaz.*, 1976, 28(9), 1570.

[2] Sitthidilokrat, N. A study on cell type of the pediatric patients with allergic rhinitis. *Thai J Pediatr.*, 2001, 40(1), 75.

[3] Jirapongsananuruk, O; Vichyanond, P. Nasal cytology in the diagnosis of allergic rhinitis in children. *Ann Allergy Asthma Immunol.*, 1998 Feb, 80(2), 165-70.

[4] Janwitayanujit, S. Anaphylaxis. Anaphylaxis. *J Med Assoc Thai.*, 2007 Jan, 90(1), 195-200.

[5] Kulthanan, K; Jiamton, S; Gorvanich, T; Pinkaew, S. Autologous serum skin test in chronic idiopathic urticaria: prevalence, correlation and clinical implications. *Asian Pac J Allergy Immunol.*, 2006 Dec, 24(4), 201-6.

[6] Wiwanitkit, V. Flux change in basophil membrane is not the main pathogenesis for hypersensitivity. *Int J Nanomedicine.*, 2007, 2(4), 663-5.

[7] Palungwachira, P; Yaguchi, H; Palungwachira, P. Electron microscopic study in a case of solitary mastocytoma. *J Med Assoc Thai.*, 2004 May, 87(5), 561-6.

[8] Tirakunwicha, S; Tulvatana, W. Disappearance of primitive basaloid cells after external irradiation in adenoid cystic carcinoma of the lacrimal gland. *J Med Assoc Thai.*, 2006 Aug, 89(8), 1318-21.

[9] Butthep, P; Bunyaratvej, A; Bhamarapravati, N. Dengue virus and endothelial cell: a related phenomenon to thrombocytopenia and granulocytopenia in dengue hemorrhagic fever. *Southeast Asian J Trop Med Public Health.*, 1993, 24 Suppl 1, 246-9.

[10] Salakij, C; Salakij, J; Apibal, S; Narkkong, NA; Chanhome, L; Rochanapat, N. Hematology, morphology, cytochemical staining, and ultrastuctural characteristics of blood cells in king cobras (Ophiophagus hannah). *Vet Clin Pathol.*, 2002, 31(3), 116-26.

[11] Wiwanitkit, V; Soogarun, S; Suwansaksri, J. Basophil and urine phenol in benzene exposed subjects. *Haema.*, 2005, 8(4), 612-614.

In: Basophil Granulocytes
Editors: Paul K. Vellis

ISBN: 978-1-60741-797-2

Chapter 7

BASOPHIL GRANULOCYTES: SMALL IN NUMBERS, MIGHTY IN IMMUNE REGULATION*

Nicolas Charles and Juan Rivera
Laboratory of Immune Cell Signaling, National Institute of Arthritis and Musculoskeletal and Skin Diseases, National Institutes of Health, Bethesda, MD, 20895, USA

Basophils are the least abundant granulocytes in the blood, barely reaching 1% of the total leukocyte population. These cells have long been ignored and their function in health and disease has remained an enigma [1]. In fact, basophils have been viewed by many as "redundant circulating mast cells". Recent advances, however, towards understanding the development of both mast cells and basophils, has allowed identification of key factors leading to the production of these cell types [2]. In humans, it is known that basophils are derived from $CD34^+$ bone marrow hematopoietic progenitor cells that yield $Fc\varepsilon RI\alpha^{hi}c\text{-}kit^-$ cells [2]. In the mouse, a common spleen progenitor leading to basophils and mast cells has been described [2]. Basophils arise from this GATA2-expressing progenitor by up regulation of the transcription factor C/EBPα. In contrast, downregulation of C/EBPα together with expression of GATA2 leads this common precursor to

* A version of this chapter was also published in Handbook of Granulocytes: Classification, Toxic Materials Produced and Pathology, edited by Reuben Hägg and Soren Kohlund, Nova Science Publishers. It was submitted for appropriate modifications in an effort to encourage wider dissemination of research.

generate mast cells [3]. However, whether these common spleen precursors lead to both kinds of cells *in vivo* remains to be established.

The lack of a basophil-deficient mouse model as well as the need for a differentiated basophil cell line has hampered studies on the role of basophils in health and diseases. Indeed, until the recent advance allowing antibody-mediated basophil depletion in mice and identification of a panel of surface marker for mouse basophils [4], the study of these cells has been limited to human patho-physiological approaches. The latter allowed the identification of basophils as important players *in vivo* in pathologies like asthma, allergic diseases and parasitic infections [1]. Like mast cells, basophils express the tetrameric form ($\alpha\beta\gamma_2$) of the high affinity receptor for immunoglobulin E (IgE) FcεRI [5]. Basophils also express a variety of other surface receptors that can lead to their activation (e.g. CD123, TLR4, FcγRIII…) [6]. FcεRI-mediated activation of basophils induces, as in mast cells, the immediate release of preformed pro-inflammatory mediators (such as histamine and proteases) and the later production and secretion of leukotrienes, cytokines and chemokines (such as IL-8, IL-4, IL-5, IL-13 and IL-10). These responses (both immediate and delayed) are potentiated with IL-3 pre-incubation of the cells [1,6].

Basophils are normally not found in tissues, but their numbers can increase both in the blood of atopic patients as well as *in situ* where allergic inflammation is occurring (airways of asthmatic patients, skin of patients with atopic dermatitis, nasal mucosa of patients with allergic rhinitis) [1,7,8]. Basophils are able to produce and release multiple mediators involved in the symptoms of both acute and chronic allergic inflammations, such as IL-4, IL-13, leukotriene C4, histamine [9]. Although basophils are not the only cells producing these products, they are the only cells found *in situ* prior to the chronic phase of allergic diseases with the ability to quickly produce significant amounts of these mediators [1]. Additionally, the basophils is the main producer of IL-4 and IL-13 in peripheral blood following the challenge of asthmatic patients [1]. However, whether basophils are necessary for the development of asthma, atopic dermatitis or allergic rhinitis remains to be determined.

The recent discovery of the basophil as a potential modulator of T_H1/T_H2 balance [10] has reawakened an interest in these cells (Figure 1). When an infection occurs, the type of pathogen prompts the immune system to determine what type of response is necessary. Both in human and mouse, it is known that some intracellular infections (virus, intracellular pathogens…) will lead naïve $CD4^+$ T cells to differentiate into T_H1 subsets, under the control of IL-12 and transcription factors such as STAT4 and T-Bet [11]. In the case of extracellular infections (such as helminth infections) or allergic disorders, naïve $CD4^+$ T cells will be driven to differentiate into T_H2 subset. The latter is controlled by IL-4 and transcription factors such as STAT6 and GATA3 [11]. While the cytokines needed for T_H1/T_H2 differentiation *in vitro* as well as *in vivo* have been extensively described both in human and mouse, the cellular compartment(s) that first initiates the production of these cytokines is poorly understood. Dendritic cells, especially around epithelia and mucosal surfaces, are involved in some *in situ* differentiation of naïve T cells into both T_H1 and T_H2 subsets (as well as T_H17 and T_{reg} subsets) [12]. However, very little was known about systemic induction of a particular T cell subset in case of systemic disorders such as atopy (T_H2 driven), autoimmunity (T_H1, T_H2 or T_H17 driven) or other chronic inflammatory disorders. Accumulating evidence tends to place the basophil compartment as a central decision maker, in particular for T_H2 differentiation (Figure 1) [13]. This hypothesis has been primarily supported by *in vitro* evidence. Indeed, it has been shown that $CD4^+$ naïve T cells, when stimulated *in vitro* in the presence of activated basophils, preferentially differentiated into IL-4-producing T cells in the absence of any exogenous addition of IL-4 [14].

Karasuyama and colleagues recently showed that basophils play an important role in IgE-mediated chronic allergic inflammation, using a murine model of this T_H2 driven disease [4,15] (Figure 1). In this particular model, depletion of basophils with an antibody specific to CD200R3 led to a reduced infiltration of inflammatory cells, such as neutrophils and eosinophils during the chronic phase of the disease [4,15]. This suggested a dominant immunomodulatory role of basophils in the development of IgE-mediated chronic allergic disease over their effector role (Figure 1). Nonetheless, basophils were also shown to play an important effector role in IgG-mediated anaphylaxis [16]. This basophil-derived PAF-mediated acute reaction was not observed if basophils were previously depleted from the mice (Figure 1) [16]. Although the relevance of IgG-mediated anaphylaxis in human patients remains to be determined, the findings demonstrate that basophils can be immunoregulatory and effector cells, depending on the stimulus involved.

The key question is what is the origin of the IL-4 that leads to the initial T_H2 differentiation in diseases such as asthma, atopic dermatitis, allergic rhinitis and others? T_H2 $CD4^+$ T cells are able to produce large amount of IL-4 which is thereafter able to maintain and amplify the T_H2 response. However, what initiates this T_H2 differentiation of $CD4^+$ T cells *in vivo* has long remained a mystery.

In 2005, Hida *et al.* [17] showed the importance of a transcription factor, interferon regulatory factor 2 (IRF2) in control of basophil proliferation *in vivo*. Indeed, $IRF2^{-/-}$ mice had a peripheral basophilia which led $CD4^+$ T cells to differentiate more easily into T_H2 subset after *in vitro* anti-CD3/anti-CD28 restimulation of the splenocytes. This was the first evidence, in the absence of any exogenous stimulus, that basophils might participate in T_H2 differentiation, albeit after *in vitro* restimulation (Figure 1) [17]. This study also suggested a key regulatory role of IL-3 in the control of basophil numbers *in vivo*. This correlates with data previously described for human basophils *in vitro* [1]. Moreover, prior *in vivo* studies from Oh et al. [14] and Lantz et al. [18] also confirmed the IL-3 importance in determining the number and function of mouse basophils. Indeed, exogenously administered IL-3 led to an increased number of basophils *in vivo* and to increased T_H2 differentiation *in vitro* after specific T cell activation [14]. Nematode infection also led to an IL-3 dependent increase in basophil numbers *in vivo*, and *in vitro* experiments showed that basophils from wild type mice responded more potently to FcεRI-mediated stimulation than *IL-3*$^{-/-}$ basophils [18].

Recent studies by Spiegl et al. [19] provide evidence of additional basophil-derived factors that can cause T_H2 differentiation. After *ex vivo* IL-3 stimulation of human basophils, these authors showed that basophils increased their synthesis of retinoic acid (RA). Basophils were the only granulocyte family member able to express retinaldehyde dehydrogenase II (RALDH2) leading to the synthesis of RA [19]. The RA produced was able to regulate gene expression in an autocrine manner, as well as in a paracrine fashion by acting on T-helper cells [19]. This finding shows that basophils can produce various factors that provide T_H2 polarization signals to the T cell compartment.

The key evidence for a direct role of basophils in T_H2 differentiation was recently provided by Sokol et al. [10] using an *in vivo* mouse model of protease-allergen challenge. Here they showed that cysteine proteases like papain or bromalein, depending on their protease activity, were able to induce protease-specific IgE production after subcutaneous immunization and drove T_H2 differentiation of naïve $CD4^+$ T cells (Figure 1) [10]. This phenomenon was dependent on basophils since *in vivo* depletion of these cells led to the lack of a specific-IgE response. Moreover, the findings showed that IL-4 and TSLP produced by the basophils after protease-induced activation was key to the response and that basophils migrated into the draining lymph nodes following challenge [10]. Thus, this study showed that, via their ability to quickly produce a robust T_H2 cytokine response (IL-4 and TSLP), *in vivo* activated basophils are able to induce T_H2 differentiation (Figure 1) [10].

However, the *in vivo* role of basophils appears to extend beyond their ability to induce T_H2 differentiation. Denzel et al. [20] recently showed that mouse basophils, through their ability to produce high levels of IL-4 and IL-6, were responsible, at least in part, for the humoral memory immune responses *in vivo*. In this elegant study, the basophils were shown to be competent antigen presenting cells, able to recognize a second exposure to the antigen through antigen-specific IgE bound to their surface expressed FcεRI (Figure 1) [20]. The FcεRI crosslinking elicited by the exposure to the antigen led basophils to migrate to the draining lymph node and to produce IL-4 and IL-6 [20]. *In vitro*, these basophil-derived cytokines induced B cells to proliferate and generate IgG1- and IgG2a producing B cells and plasma cells via a $CD4^+$ T helper cells and through a contact (CD40/CD40L) dependent mechanism. Moreover, *in vitro* activated basophils increased the "B-cell helper" T_H2 phenotype of $CD4^+$ T cells (Figure 1) [20]. However, these *in vitro* activated basophils were also able to induce IgG2a production, showing that in this model basophils were able to induce both T_H1 and T_H2 humoral responses [20]. If basophils, B cells and T cells interact *in vivo*, the site in which this occurs is still to be defined. Furthermore, how the antigen presenting basophil in the blood, spleen or bone marrow is recruited to the site of interaction remains unclear.

While the aforementioned studies provide convincing evidence of a key role for the basophil in T_H2 differentiation and in enhancing humoral responses, our knowledge of the molecular factors controlling the basophils ability to induce these responses has been limited. We recently showed, however, that basophils from mice deficient in Lyn kinase caused a strong constitutive T_H2 skewing *in vivo*. Lyn is a member of the Src family protein tyrosine kinases (SrcPTK). It is known to function as a positive and a negative regulator of the FcεRI signaling in mast cells [21]. Lyn is expressed in most haematopoietic cells but not in T cells. Interestingly, *lyn*$^{-/-}$ mice develop an early life onset of an atopic-like phenotype. They have increased serum IgE levels [22], are more susceptible to asthma challenge [23] as well as IgE-mediated anaphylaxis [22]. These mice also have a profoundly altered B cell compartment [24,25], with B cells being hypersensitive to IL-4 and CD40/CD40L stimulation [25]. In these mice, the number of circulating B cells is dramatically reduced, but the life span of B and plasma cells are increased [25]. In order to understand which cells and factors were responsible for the atopic-like phenotype, we analyzed the basophil compartment in these mice. Basophil numbers and FcεRI-mediated IL-4 production per cell were dramatically increased in *lyn*$^{-/-}$ mice [26]. These mice developed an early and inappropriate onset of T_H2 responses in the papain-induced model of T_H2 response [10,26]. This phenotype was linked to the constitutive T_H2 skewing of the T cell compartment of the *lyn*$^{-/-}$ mice [26]. In the absence of any exogenous challenge, $CD4^+$ T cells from the spleen of *lyn*$^{-/-}$ mice were T_H2 skewed. This T_H2 skewing was mast cell-independent but dependent on IgE, IL-4 and basophils [26]. The *in vivo* depletion of basophils in *lyn*$^{-/-}$ mice completely reversed the T_H2

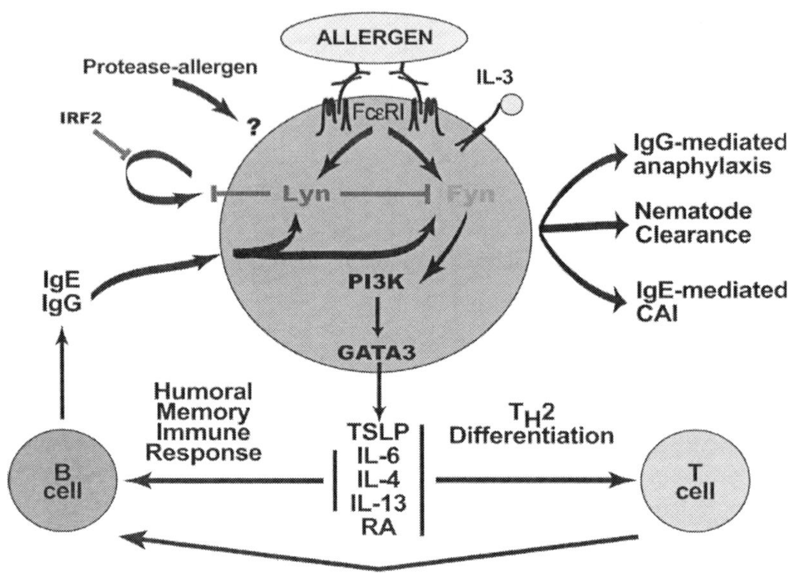

Figu skewing of the mice. The use of mast cell-deficient mice and mast cell/lyn-double deficient mice also demonstrated that the *in vivo* FcεRI engagement of basophils alone (since mast cells were not present), results in the induction of cytokines that selectively induced T_H2, and not T_H1, differentiation of the $CD4^+$ T cells [26]. The ability of the basophil to produce IL-4 was found to be associated with the induction of the transcription factor GATA3 and required the activation of phosphatidylinositol 3-OH kinase. Thus, Lyn kinase is a key factor in basophils that controls the onset and extent of T_H2 differentiation through its ability to control basophil proliferation as well as GATA3 induction and IL-4 production in FcεRI-activated basophils (Figure 1) [26].re 1. Physiological and Pathological Role of Basophils: Control of T_H2 Differentiation and Humoral Responses. It is now established both *in vitro* and *in vivo*, that basophils can induce T_H2 differentiation via their cytokine production. Stimulation of basophils through protease-allergen, FcεRI crosslinking and/or IL-3 receptor engagement leads to the release of mediators like: IL-4, IL-6, IL-13, TSLP and retinoic acid (RA). The release of these mediators *in vivo* enhances humoral memory immune responses and initiates T_H2 differentiation resulting in IgE and IgG specific responses. The basophils can then manifest their key immunoregulatory and effector roles in IgG-mediated anaphylaxis, IgE-mediated chronic allergic inflammation and nematode clearance. The FcεRI-dependent stimulation of basophils, causes enhanced Lyn kinase activity which down-regulates the Fyn kinase whose activity is key for phosphatidylinositol 3-OH kinase (PI3K) activation, which controls the levels of GATA3 expressed in the basophil. GATA3 has been shown to be a determinant of IL-4 production and secretion, and thus Lyn kinase controls the T_H2 differentiation of the $CD4^+$ T cell subset by suppressing GATA3 expression and IL-4 production in basophils. In addition, Lyn kinase and the transcription factor IRF2 suppress the proliferation of peripheral basophils. Thus, this also contributes to the dampening of the overall IL-4 production *in vivo*. Dysregulation of these controls is likely to lead to disease as manifest in *lyn*$^{-/-}$ mice.

skewing of the mice. The use of mast cell-deficient mice and mast cell/lyn-double deficient mice also demonstrated that the *in vivo* FcεRI engagement of basophils alone (since mast cells were not present), results in the induction of cytokines that selectively induced T_H2, and not T_H1, differentiation of the $CD4^+$ T cells [26]. The ability of the basophil to produce IL-4 was found to be associated with the induction of the transcription factor GATA3 and required the activation of phosphatidylinositol 3-OH kinase. Thus, Lyn kinase is a key factor in basophils that controls the onset and extent of T_H2 differentiation through its ability to control basophil proliferation as well as GATA3 induction and IL-4 production in FcεRI-activated basophils (Figure 1) [26].Collectively, the recent studies have brought new knowledge on the role of basophils *in vivo*. However, a number of questions still remain. First, what are the factors essential for basophil lineage ontogeny? This is particularly important if we wish to realize the therapeutic potential of manipulating this compartment *in vivo*. Indeed, the identification, in human, of such early and late, multipotent or specific precursors, would allow understanding of the conditions needed for controlling basophil numbers *in vivo* in pathologies such as asthma and/or atopic disorders. This knowledge might allow developing specific strategies that control T_H2 differentiation.

To what extent are the recent findings in mouse models clinically applicable to human pathology? Is the basophil the key cell involved in the development of chronic T_H2-mediated diseases? Further studies are necessary to answer these questions. For instance, basophils can be activated by proteases contained in house dust mite [1], by IgE specific to parasite-derived proteins such as schistozomal egg proteins [6], and by IgE and IgG containing circulating immune complexes [27]. Would the temporary depletion of basophils in humans be a good strategy to abrogate the unwanted chronic inflammation in such diseases as asthma, chronic allergic inflammation or parasitic infections? On the other hand, given the demonstrated key immune regulatory role of basophils, common treatments used in chronic inflammatory disorders, such as steroids, which decrease both the numbers and reactivity of basophils [28,29], could lead to the complete loss of endogenous control of immune regulation. Thus, the renewed interest in basophil biology is likely to lead to an increased understanding of how and when this cell type controls immune responses, which may have value in therapeutic and preventive care.

REFERENCES

[1] Schroeder, JT; Kagey-Sobotka, A; Lichtenstein, LM. The role of the basophil in allergic inflammation. *Allergy*, 1995, 50, 463-472.

[2] Iwasaki, H; Akashi, K. Myeloid lineage commitment from the hematopoietic stem cell. *Immunity*, 2007, 26, 726-740.

[3] Arinobu, Y; Iwasaki, H; Gurish, MF; Mizuno, S; Shigematsu, H; Ozawa, H; Tenen, DG; Austen, KF; Akashi, K. Developmental checkpoints of the basophil/mast cell lineages in adult murine hematopoiesis. *Proceedings of the National Academy of Sciences of the United States*, 2005, 102, 18105-18110.

[4] Mukai, K; Matsuoka, K; Taya, C; Suzuki, H; Yokozeki, H; Nishioka, K; Hirokawa, K; Etori, M; Yamashita, M; Kubota, T; Minegishi, Y; Yonekawa, H; Karasuyama, H. Basophils play a critical role in the development of IgE-mediated chronic allergic inflammation independently of T cells and mast cells. *Immunity*, 2005, 23, 191-202.

[5] Kinet, JP. The high-affinity IgE receptor (Fc epsilon RI): from physiology to pathology. *Annual Review of Immunology*, 1999, 17, 931-972.

[6] Min, B. Basophils: what they 'can do' versus what they 'actually do'. *Nature Immunology*, 2008, 9, 1333-1339.

[7] Howarth, PH; Salagean, M; Dokic, D. Allergic rhinitis: not purely a histamine-related disease. *Allergy*, 2000, 55 Suppl 64, 7-16.

[8] Marone, G; Triggiani, M; de Paulis, A. Mast cells and basophils: friends as well as foes in bronchial asthma? *Trends in Immunology*, 2005, 26, 25-31.

[9] Min, B; Paul, WE. Basophils and type 2 immunity. *Current Opinion in Hematology*, 2008, 15, 59-63.

[10] Sokol, CL; Barton, GM; Farr, AG; Medzhitov, R. A mechanism for the initiation of allergen-induced T helper type 2 responses. *Nature Immunology*, 2008, 9, 310-318.

[11] Chen, Z; O'Shea, JJ. Th17 cells: a new fate for differentiating helper T cells. *Immunologic Research*, 2008, 41, 87-102.

[12] Reschner, A; Hubert, P; Delvenne, P; Boniver, J; Jacobs, N. Innate lymphocyte and dendritic cell cross-talk: a key factor in the regulation of the immune response. *Clinical and Experimental Immunology*, 2008, 152, 219-226.

[13] Karasuyama, H; Mukai, K; Tsujimura, Y; Obata, K. Newly discovered roles for basophils: a neglected minority gains new respect. *Nature Reviews in Immunology*, 2009, 9, 9-13.

[14] Oh, K; Shen, T; Le Gros, G; Min, B. Induction of Th2 type immunity in a

mouse system reveals a novel immunoregulatory role of basophils. *Blood*, 2007, 109, 2921-2927.

[15] Obata, K; Mukai, K; Tsujimura, Y; Ishiwata, K; Kawano, Y; Minegishi, Y; Watanabe, N; Karasuyama, H. Basophils are essential initiators of a novel type of chronic allergic inflammation. *Blood*, 2007, 110, 913-920.

[16] Tsujimura, Y; Obata, K; Mukai, K; Shindou, H; Yoshida, M; Nishikado, H; Kawano, Y; Minegishi, Y; Shimizu, T; Karasuyama, H. Basophils play a pivotal role in immunoglobulin-G-mediated but not immunoglobulin-E-mediated systemic anaphylaxis. *Immunity*, 2008, 28, 581-589.

[17] Hida, S; Tadachi, M; Saito, T; Taki, S. Negative control of basophil expansion by IRF-2 critical for the regulation of Th1/Th2 balance. *Blood*, 2005, 106, 2011-2017.

[18] Lantz, CS; Min, B; Tsai, M; Chatterjea, D; Dranoff, G; Galli, SJ. IL-3 is required for increases in blood basophils in nematode infection in mice and can enhance IgE-dependent IL-4 production by basophils in vitro. *Laboratory Investigation*, 2008, 88, 1134-1142.

[19] Spiegl, N; Didichenko, S; McCaffery, P; Langen, H; Dahinden, CA. Human basophils activated by mast cell-derived IL-3 express retinaldehyde dehydrogenase-II and produce the immunoregulatory mediator retinoic acid. *Blood*, 2008, 112, 3762-3771.

[20] Denzel, A; Maus, UA; Rodriguez Gomez, M; Moll, C; Niedermeier, M; Winter, C; Maus, R; Hollingshead, S; Briles, DE; Kunz-Schughart, LA; Talke, Y; Mack, M. Basophils enhance immunological memory responses. *Nature Immunology*, 2008, 9, 733-742.

[21] Rivera, J; Gilfillan, AM. Molecular regulation of mast cell activation. *The Journal of Allergy and Clinical Immunology*, 2006, 117, 1214-1225.

[22] Odom, S; Gomez, G; Kovarova, M; Furumoto, Y; Ryan, JJ; Wright, HV; Gonzalez-Espinosa, C; Hibbs, ML; Harder, KW; Rivera, J. Negative regulation of immunoglobulin E-dependent allergic responses by Lyn kinase. *The Journal of Experimental Medicine*, 2004, 199, 1491-1502.

[23] Beavitt, SJ; Harder, KW; Kemp, JM; Jones, J; Quilici, C; Casagranda, F; Lam, E; Turner, D; Brennan, S; Sly, PD; Tarlinton, DM; Anderson, GP; Hibbs, ML. Lyn-deficient mice develop severe, persistent asthma: Lyn is a critical negative regulator of Th2 immunity. *The Journal of Immunology*, 2005, 175, 1867-1875.

[24] Hibbs, ML; Tarlinton, DM; Armes, J; Grail, D; Hodgson, G; Maglitto, R; Stacker, SA; Dunn, AR. Multiple defects in the immune system of Lyn-deficient mice, culminating in autoimmune disease. *Cell*, 1995, 83, 301-311.

[25] Nishizumi, H; Taniuchi, I; Yamanashi, Y; Kitamura, D; Ilic, D; Mori, S;

Watanabe, T; Yamamoto, T. Impaired proliferation of peripheral B cells and indication of autoimmune disease in lyn-deficient mice. *Immunity*, 1995, 3, 549-560.

[26] Charles, N; Watford, WT; Ramos, HL; Hellman, L; Oettgen, HC; Gomez, G; Ryan, JJ; O'Shea, JJ; Rivera, J. Lyn Kinase Controls Basophil GATA-3 Transcription Factor Expression and Induction of Th2 Cell Differentiation.. *Immunity*, 2009, 30, 533-543.

[27] Mukai, K; Obata, K; Tsujimura, Y; Karasuyama, H. New Insights into the Roles for Basophils in Acute and Chronic Allergy. *Allergology International*, 2009, 58, 11-19.

[28] Schleimer, RP. The mechanisms of antiinflammatory steroid action in allergic diseases. *Annual Review of Pharmacology and Toxicology*, 1985, 25, 381-412.

[29] Schleimer, RP; Davidson, DA; Peters, SP; Lichtenstein, LM. Inhibition of human basophil leukotriene release by antiinflammatory steroids. *International Archives of Allergy and Applied Immunology*, 1985, 77, 241-243.

In: Basophil Granulocytes
Editor: Paul K. Vellis

ISBN: 978-1-60741-797-2

Chapter 8

C-Reactive Protein: FCR Receptor-Mediated Effects on Human Peripheral Blood Basophils in Vitro[*]

Peter G. Nazarov[†1]***, Anastasia P. Pronina***[2]
and Andrey S. Trulioff[2]

[1] Professor, Head, Laboratory of General Immunology,
Department of Imminology, Institute of Experimental Medicine,
12 Acad. Pavlov Street, 197376, St. Petersburg, Russia.
[2] Department of Immunology, Institute of Experimental Medicine,
St. Petersburg, Russia.

Introduction

Described by Paul Ehrlich in 1879, basophilic granulocytes (basophils) remain among the least studied white blood cells. A rather small number of works on them (compared to other leukocyte types) was attributable to their extremely

---

[*] A version of this chapter was also published in C-Reactive Protein: New Research, edited by Satoshi Nagasawa, Nova Science Publishers. It was submitted for appropriate modifications in an effort to encourage wider dissemination of research.

[†] Corresponding author: Tel.: +7 812 234 1669 (work), +7 812 543 5214 (home); Fax: +7 812 234 9489. Email: peter_nazarov@mail.ru, pnazarov@pn4093.spb.edu (P.G. Nazarov)

low content in the blood (0,5-1% of nucleated blood cells), but not to lack of interest by scientists. The number and morphology of basophils are somewhat different in different animals, but they occur in all classes of vertebrates. It demonstrates the importance of these cells for protective systems. Basophils are involved in allergic reactions and anti-parasitic protection. Moreover, due to cytokine secretion, in particular IL-4 and IL-13, they became the focus of attention as a cell type implicated in the immune response switch toward Th2 [Falcone et al., 2006]. The relationship of basophils with the acute phase of inflammation factors, in particular, with C-reactive protein (CRP), are poorly elucidated.

Elevated expression of CRP, a pentraxin and known marker of acute phase of inflammation, accompanies the onset of any form of immune response. However, CRP influence on the intensity and specificity of immune reactions is not clearly understood. It has been known that a good correlation exists between CRP production at early stage of immunization and subsequent antibody production in rabbits. It has been also shown that CRP could influence the fine specificity of immune responses to antigens which are its ligands. CRP could decrease protective antibody production against pneumococcal cell wall phosphorylcholine (PC). PC binding by CRP leads to the opsonization of pneumococcci, complement and phagocytosis activation [Nakayama S. et al., 1984; Horowitz J. et al., 1987; Szalai A.J. et al., 1995]. There are numerous data concerning stimulating effects of CRP on phagocytic activity and other functions of macrophages and neutrophils. CRP has been shown to depend for its biological function on cellular Fcγ receptors (FcgRs). FcgRI is a high-affinity IgG receptor expressed on myeloid cells that is up-regulated during inflammation [Van de Winkel J.G.J., Anderson,C.L., 1991; Van de Winkel J.G.J., Capel P.J., 1993; Beekman J.M. et al., 2004]. However, the major receptor for CRP on leukocytes is low affinity FcgRII [Bharadwaj D. et al., 1999].

The least understood is the character of the pentraxin influence upon the immediate type hypersensitivity reactions. There are few data suggesting suppressive effects of CRP on allergy development. Thus, it has been shown that the levels of CRP and IgE were negatively correlated in the population of asbestos workers [Lange A. et al., 1995]. However these data do not answer all the questions concerning participation of pentraxins and especially CRP in allergic reactions of acquired immunity.

We have shown in guinea pigs with experimental anaphylaxis that the injection of purified CRP prior to antigen attenuated anaphylactic reaction to a significantly greater extent than the injection of normal human IgG [Nezhinskaya G.I., Nazarov P.G., Evdokimova N.R., Losev N.A., Sapronov N.S., 2004;

Nazarov P.G. et al., 2005]. We have also shown that purified human CRP diminished effects of acetylcholine (Ach) on the vascular tone and the heart rate of rats in vivo. In vitro CRP inhibited breakdown of Ach by acetylcholinesterase (AchE) while did not interact with AchE itself. CRP did not modify the cardiovascular effects of adenosine, another vasorelaxant. These data suggest that Ach sharing structural features with phosphorylcholine, the classical ligand of CRP, can also be a ligand of the pentraxin [Nazarov P.G. et al., 2006, 2007]. CRP might act on Ach itself by capturing it or on cellular Ach receptors by blocking them or modifying their function. Ameliorating effect of CRP on the development of allergic sensitization and anaphylactic shock, we have described, could be due to the fact that CRP binds Ach and reduces its proinflammatory effects. Moreover, anti-anaphylactic effect of CRP could also be linked to the action of the pentraxin on basophils and mast cells, but this question is not good elucidated in the literature.

Blood levels of CRP are positively correlated with cardiovascular disease risk and endothelial dysfunction. An inhibitory effect of CRP on endothelial function has also been shown by other authors who explained it by the decrease in NO production due to inhibition of PP2A phosphatase and down-regulation of endothelial NO synthase [Devaraj S. et al., 2005; Liang Y.-J. et al., 2006]. CRP antagonism of eNOS was reported to be attributable to blunted eNOS phosphorylation at Ser1179. This was supported by the data that in the absence of active PP2A (due to inactivation by short-interference RNA) CRP had no effect on eNOS.

Our previous data are inconsistent with the conclusion that the eNOS down-regulation and NO production impairment by CRP are major reasons of endothelial dysfunction caused by the pentraxin. Our data have indicated that CRP had no effect on vascular relaxation caused by adenosine, the known NO inducer [Nazarov P.G. et al., 2007]. However, whether CRP has direct actions on endothelial cells and the mechanisms underlying such actions are unknown.

One of the mechanisms of the anti-anaphylactogenic effect of CRP shown by us earlier could be associated with the pentraxin inhibitory action on tissue mast cells and blood basophils responsible for the release of vasoactive mediators and the response of vessels and bronchi in allergic reactions of immediate type. Mast cells and basophils are activated by IgE antibodies and relevant antigen through IgE receptors (FceRI). The receptors FcgRI and FcgRIII, binding IgG subclasses (IgG1, IgG2, IgG3), are also activators of these cells, whereas FcgRIIb, on the contrary, inhibits cell activity and cancels activation signals of other receptors. Such pentraxins as CRP and serum amyloid P-component (SAP) known to be actively expressed in acute phase of inflammations influence the immunocompetent cell activity also through Fc-gamma R. The role of pentraxins in activation of mast cells and basophils is poorly defined. The cholinergic regulation of mast cell and basophil activity is also poorly understood, in particular in view of recent attention to autonomous, non-neuronal cholinergic system of immunocompetent cells which includes such elements as Ach synthesis, Ach destruction, and Ach receptors of muscarinic and nicotinic type (mAchR and nAchR) [Kawashima K., Fujii, T., 2000; Kirkpatrick, C.J. et al., 2003; Wessler I. et al., 2003]

The aim of the study was to investigate the ability of C-reactive protein to activate human blood basophils in vitro and to elucidate the role of autonomic cholinergic system elements (Ach and AchRs) in modulation of the pentraxin effect on basophils. Basophil functional activity activated by CRP or other inductors was investigated in the presence or absence of cholinergic blockers, such as mAchR and nAchR antagonists.

Materials and Methods

Basophils Isolation

Basophils were isolated from heparinized peripheral blood of healthy donors. For each separation approximately 30-40 ml of venous blood from one donor were used. The first step of the protocol was the preparation of leukocyte mixture by spontaneous red blood cells sedimentation (approximately 30 min at 37 °C) or by gradient centrifugation in Ficoll-Paque. Human Basophil Isolation Kits II purchased from Miltenyi Biotec was used for basophil separagtion. Separation protocol was based on negative selection of basophils. All leukocytes exept basophils were first treated with a cocktail of biotinilated monoclonal antibodies against CD3, CD4, CD7, CD14, CD15, CD16, CD36, CD45RA, HLA-DR and CD235a and then with anti-biotin monoclonal antibody conjugated with magnetic microbeads. Magnetic particles with attached cells were then removed using MACS® Column and magnetic field in MACS Separator. This approach provided a highly enriched basophil fraction. The cells eluted from the column were qualified as intanct because they did not contact with any antibody.

Isolated basophils were analysed by flow cytometry on Epics Altra Cell Sorter (Beckman-Coulter) using FITC-labelled monoclonal antibodies against CD63 and CD203c. CD63 (gp53, or lyzozyme-associated membrane protein, LAMP-3) is a transmembrane protein expressed not only by basophils, but also by tissue mast cells, macrophages and platelets [Nieuwenhuis H.K. et al.,1987; Metzelaar M.J. et al., 1991; Grutzkau A. et al., 2004]. In resting basophils and mast cells CD63 is hidden inside the cell being a component of cytoplasmic granules, but after activation appears on the cell surface. On the IgE-activated basophils, during their degranulation, CD63 is expressed in high density due to fusion of the granules with cell surface membrane [Knol E.F., 1991]. CD203c (neural cell surface differentiation antigen E-NPP3; PD-Ib, B10, gp130RB13-6) belongs to a multigene family of ectonucleotidepyrophosphatases/ phosphodiesterases (E-NPPs), that includes also E-NPP1 (PC-1, PDNP-1) and E-NPP2 (PD-Ia, PDNP2, autotoxin). In peripheral blood, CD203c is expressed exclusively on the surface of basophils [Bühring H.J. et al., 2004]. In contrast to CD63, CD203c is expressed on the surface of resting basophils and after their activation the expression of CD203c is only slightly enhanced; that's why this marker is less prominent than CD63.

Reagents

Armine, an irreversible AchE inhibitor, was used as 0.01% solution. Final concentration in cell suspension was 2 μg/ml or 0.2 μg/ml. Hexamethonium, a nicotinic AchR antagonist, was used as a sterile 2.5% solution for injections; final concentration was 4.16 mg/ml. Methacine (sterile 0.01% solution for injections); final concentrations were 0.017% (0.17 mg/ml) or 0.0017% (0.017 mg/ml). Armine, hexamethonium and methacine were purchased from Kazan Pharmaceuticals (Russia), concanavalin A (ConA) from Pharmacia Fine Chemicals (Sweden), compound 48/80, carbachol (carbamylcholine chloride) and histamine dihydrochloride and – from Sigma-Aldrich. Concentrations of carbachol and ConA are indicated under "Results and Discussion" section.

Human C-reactive protein (CRP) was obtained from MB-Biochemicals, human serum amyloid P-component (SAP) – from Calbiochem, histamine diphosphate salt – from Sigma-Aldrich, normal human immunoglobulin (IgG) – from Institute Pesteur (St. Petersburg, Russia), anti-CD16 monoclonal antibody (ICO-116) – from MedBioSpektr (Moscow, Russia). Posphate buffered salt solution (PBS) contained 0.14 M sodium chloride in bidistilled water buffered with a mixture of phosphate salts, pH 7.2. Heat aggregated IgG (aIgG) was prepared by heating IgG solution at a concentration of 10 mg/ml in PBS in a water bath at 63°C for 10 min. Prior to use, solutions of commercially available CRP and SAP were dialyzed overnight at 4°C against large excess of sterile PBS using

Slide-A-Lyzer Cassettes (Pierce) to remove NaN_3. LPS of *Salmonella typhi* was purchased from the Institute of Vaccines and Sera (St. Petersburg, Russia).

Cell Incubations and Assessment of Basophil Response

Isolated and washed blood cells were suspended in PBS and put into eppendorf polypropylene tubes, 10^5 cells per tube in 800 μl on ice. CRP (or aIgG, or SAP) was added in appropriate volume to get final concentrations indicated under the "Result and Discussion" section. Cholinergic reagents (nAChR or mAChR antagonists, AchE inhibitor, or PBS for control tubes) were added immediately after. Total volume was 960 μl, including cells, a protein stimulant, and a cholinergic agent. Tubes were placed into incubator and incubated at 37°C for 30 min. Then the tubes were cooled on ice and centrifuged at 400 g for 10 min at 4°C. Supernatants were collected and frozen at -20°C for histamine determination. Cell pellets were frozen separately for total histamine content determination. The Shore method with orthoftalic aldehyde and fluorometric detection [Fujimoto T. et al., 2003] was used. Triplicates of each sample were measured in a microplate fluorometer (Victor 5, Wallac) at wavelengths of 340 (excitation) and 450 nm (emission). The numbers of experiments are as indicated in Figures below.

Concentration of histamine in the supernatants was estimated from the fluorometer readings using a linear calibration plot built in each experiment with standard histamine solutions. The data on histamine secretion are presented as nanograms of histamine recovered in cell supernatants after incubation with stimulants (ng/ml per 10^5 cells), as well as stimulation indexes which were calculated as ratio of histamine (in ng/ml) released in the presence of the agents to spontaneous histamine release (ng/ml) in PBS (the latter was taken to be 1.0).

Basophil Desensitizsation and Restimulation in Vitro

Leukocyte fraction isolated from fresh healthy donor blood by spontaneous erythrocyte sedimentation was used in these experiments. The cells were suspended in RPMI 1640 tissue culture medium (Biolot, St. Petersburg, Russia) supplemented with 10% fetal calf serum (FCS, Flow Laboratories, UK) of L-glutamine (2 mmol/L) and gentamicin sulfate (50 mg/ml) and cultivated in 24-well plastic plates (Linbro) (5×10^5 per well) in CO2-incubator (Flow) at 37°C for 48 hr. The stimulations of the cultures were performed 2 or 3 times, once a day. The first stimulation was at the beginning of the cultivation. For this purpose,

CRP, aIgG, ConA, Cch, or PBS were added to the cultures for 40 min at 37°C. Thereafter the cells were sedimented, supernatants removed for histamine determination, and the cell pellets washed thrice with PBS, resuspended in fresh tissue culture medium (its composition see above) and allowed to cultivate further at 37°C. Washing fluids after last wash were analyzed for histamine content to control the completeness of washing. On the next day, the cell cultures were restimulated with the same stimulants by adding them again to the culture medium for 40 min at 37°C. Histamine concentration in test supernatants and control washing fluids was determine by Shore method.

Statistical Analysis

Values presented in Figures are means (M) ± standard errors of mean (SEM). Student *t* test was used to assess differences between groups, and significance was set at $p<0.05$.Stimulation indices were calculated as ratio of histamine quantity (in ng/ml) released in the presence or each agent to spontaneous histamine release (ng/ml) in PBS (taken to be 1.0).

Results and Discussion

Basophil Responses to Human Aggregated IgG, C-Reactive Protein, and Fcgriii Cross-Linking

Aggregated IgG is a well-known ligand for Fcg receptors. The data on histamine release from healthy donor blood basophils incubated in vitro with human aIgG, CRP, SAP and monoclonal anti-CD16 antibody are shown in Figure 1. In our experiments, both normal human IgG and normal human IgG aggregated by heating at 63 C caused significant increase in histamine release. Heat aggregated IgG was slightly more active in basophil stimulating compared to non-aggregated IgG preparation. The mean stimulation index for aIgG (100 μg/ml) was 8.5±1.2 (n=3) whereas for non-aggregated 5.6±1.1 (n=3). Difference between the effects of IgG and aIgG was not significant ($p>0.05$).

Human CRP induced a dose-dependent basophil responses with mean stimulation indexes being 0.94±0.02 (n=3) at 10 μg/ml, 3.56±0.06 (n=6) at 50 μg/ml, and 1,67±0.02 (n=3) at 100 μg/ml. So CRP enhanced histamine secretion in a dose-dependent way with a maximum at 50 μg/ml. SAP caused a weak stimulation of basophil secretion (mean stimulation index with SAP was 1.215±0.02 (n=3), whereas anti-CD16 antibody caused apparent stimulatory effect ($p<0.05$). These data indicate that cross-linking of low affinity FcgRIII activates basophils to enhance histamine release.

Our experimental data show (Figure 1) that the effect of human pentraxin CRP upon histamine liberation by basophils was similar to the effects of other agents capable of binding to FcgR, such as aggregated IgG or anti-FcgRIII monoclonal antibody. All of them enhanced histamine liberation from basophils.

A moderate but statistically significant histamine release was stimulated by incubation with *S. typhi* LPS (25 μg/ml) (Figure 1). The most effective histamine liberator was the compound 48/80 stimulated the mean histamine release at a level of about 73% of the total histamine content in the cells (Figure 1).

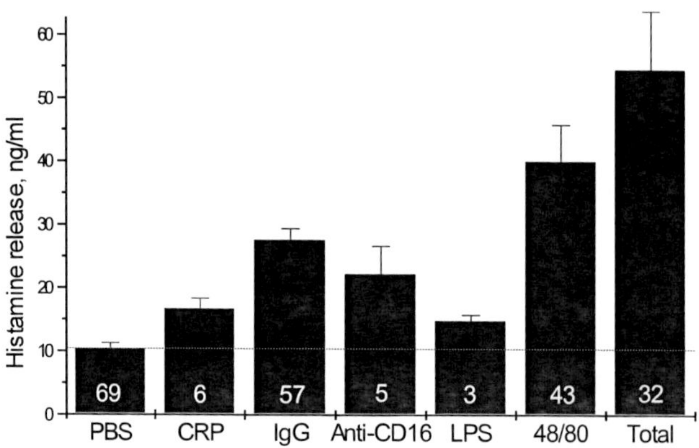

Numbers at the bottom of the columns represent the number of experiments.

Figure 1. Histamine release from normal human blood basophils incubated *in vitro* with human IgG, CRP, and anti-CD16 antibody.

2. Basophil Responses to Cholinergic Drugs

A list of used cholinergic drugs included: carbachol, a non-metabolizable chemically related analog of Ach that does not readily enter cells; hexamethonium, a nicotinic Ach receptor antagonist; methacine, a muscarinic AchR antagonist; and armine, an irreversible inhibitor of AchE.

Carbachol displayed a two-zone effect depending on its concentration in the medium. Its dose-response curve is shown on Figure 2. At high doses (0.15-1.5 mg/ml) carbachol was suppressive and inhibited histamine response below the level of spontaneous release observed with PBS. At a lower concentration it showed stimulatory effect on basophils and elevated histamine secretion twofold compared to control level. The carbachol dose of 15 μg/ml was selected as the optimal for further experiments.

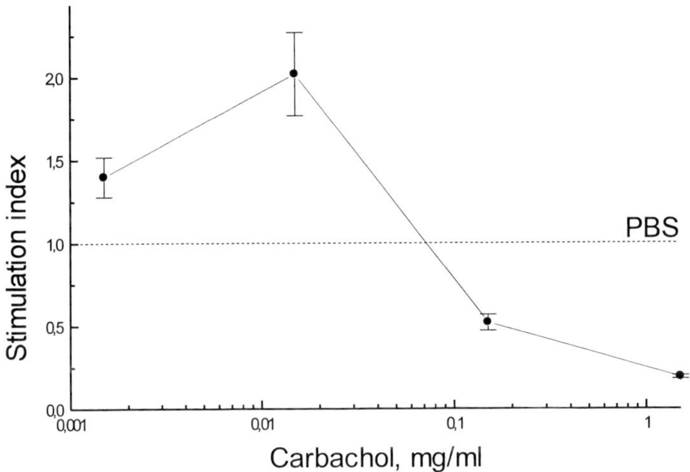

Each point is a mean of three tests ± SEM.

Figure 2. Human blood basophil responses to carbachol. A dose – response curve.

Figure 3 shows the effects of four cholinergic drugs on histamine release from human basophils *in vitro*.

Of interest is the effect of armine on basophil secretion. Under its action a complete block of AchE occurs leading to complete termination of Ach breakdown and accumulation of undestroyed, active Ach. So, in the presence of armine the excessive formation of Ach inside the cells and in their environment should occur. As can be seen from Figure 3, the increase in basophil histamine secretion in the presence of armine did occur, like that observed under the direct action of Ach-related carbachol ($p>0.05$). We have tested two concentrations of armine, 0.2 and 2 μg/ml. Figure 3 shows the effect of only one of them (2 μg/ml); data on the effect of 0.2 μg/ml are not shown. Difference between the effects of these two doses of armine was not statistically significant ($p>0.05$).

It can also be seen from Figure 3 that the blockade of nicotinic AchRs with hexamethonium at a final dose of 4.17 mg/ml resulted in a 1.5-fold stimulation of histamine secretion. Hexamethonium stimulatory effect was reproducible and statistically significant ($p<0.05$). Hexamethonium is largely a nAChR blocker and is supposed to occupy only the nicotinic part of the cell AchR pool. Its enhancing effect on basophil activity might be due to a compensatory increase in cholinergic signaling through an antagonistic pathway associated with mAChRs.

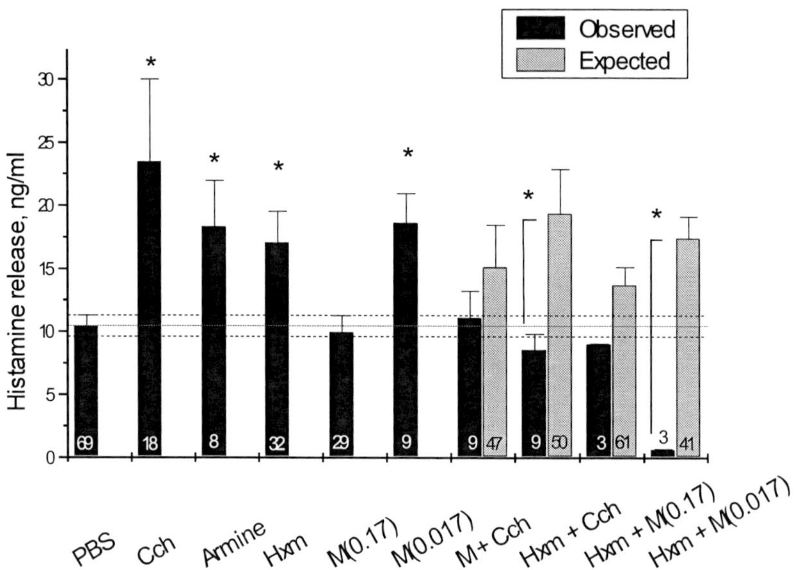

Numbers at the bottom of the columns represent the number of experiments.

Figure 3. Human blood basophil responses to hexamethonium, methacine, armine and their combinations.

Similar stimulating effects were seen with methacine. This muscarinic antagonist produced an enhancement of histamine secretion in low dose of 0.017 mg/ml (Figure 3) while the higher dose of 0.17 mg/ml showed no effect on basophil secretion (Figure 3). This result may be attributable to the reasons mentioned above for hexamethonium. Occupation of muscarinic subpopulation of cellular AchRs by methacine could facilitate a compensatory Ach flow via alternative nAChRs.

Thus, carbachol and armine stimulated histamine secretion by normal human blood basophils. Carbachol is known to be similar to Ach by its activity, while differs from it by the resistance to AchE action. Based on the effect of carbachol on basophils one can get idea of the effect of Ach. Armine blocks the AchE and stops endogenous Ach destruction, thus increasing the influence of Ach on cells. So the increase in histamine secretion induced by either carbachol or armine may be attributable to Ach activity. Therefore, it can be concluded that the effects of carbachol and armine are associated with Ach activity. Conceivably both carbachol and armine might mediate their effects through either nAChR- or mAChR-dependent pathways.

By combined application of carbachol with nAChR or mAChR antagonist we have analyzed the question of which type of AchRs mediated the action of exogenic carbachol and endogenous Ach accumulating under the AchE inhibition. As to carbachol, some data suggest that it is a selective agonist of muscarinic AchRs [Oishi K. et al., 2004]. In contrast to these data, however, another evidence exists indicating the absence of selectivity of carbachol [Xiao Y. and Kellar K.J., 2004]. In addition, we could not find published data on the nAChR or mAChR selectivity of endogenous Ach accumulating under the AchE inhibition.

Statistical comparisons of the observed combined effects of the drugs were made with the expected (rated) values calculated from their separate effects. Figure 3 shows separate effects of drugs, their real combined effects, and the calculated values of combined effects referred to as “expected”, or “rated”.

As can be seen, incubation of the cells with carbachol together with methacine induced much lower histamine release than expected. Similarly, combination of carbachol with hexamethonium also provided significantly less pronounced effect of histamine release than could be expected based on their separate effects ($p<0.05$, Fifure 3). The use of calculated (“expected”) values rather than real separate controls seemed to be more correct for comparison purposes since such calculated values would represent more weighted average estimates of any joint effect than do any of the source control values. So, the comparisons with the “expected” effects indicated that the blockade of either nAChRs or mAChRs resulted in cessation of stimulatory effect of carbachol on histamine release from blood basophils in vitro. The inhibitory effect of nAChR antagonist hexamethonium was even more pronounced than the effect of muscarinic antagonist methacine. These data do not fit the concept that carbachol has muscarinic selectivity. Our results suggest that this Ach analog can provide its activation signals both to nAChRs and mAChRs.

The last point from Figure 3 to be discussed is the combined effect of two AchR antagonists. In these experiments hexamethonium was used in one dose (4.17 mg/ml), while methacine in two (0.017 and 0.17 mg/ml). Both hexamethonium and methacine when acting singly enhanced basophil histamine secretion (Figure 3). Treatment of the cells with their mixture led to the cessation or even abrogation of stimulating effect. It suggests that the withdrawn of any cholinergic pathway (either muscarinic or nicotinic) results in compensatory enhancement of signaling through another one. The blockade of both AchR types cuts of signaling, and Ach can not further activate the cells. As can be seen from Figure 3, a combination of hexamethonium with methacine was less stimulatory than expected. The most suppressive was the lower dose of methacine (0.017 mg/ml). In combination with hexamethonium it suppressed basophil activity significantly, much below background level ($p<0.05$). The reason of such profound suppression may be that the low dose of methacine selects and trigger a high affinity subset of mAChRs that has a suppressive function in basophil physiology control, and this results in the activation of a suppressive mechanism like heterologous desensitization.

3. Basophil Responses to Aggregated IgG and Cholinergic Drugs

Aggregated IgG was used as the classical ligand for FcgRs. As can bee seen from Figure 4, aIgG was able to induce histamine secretion by normal blood basophils from healthy donors. At a concentration of 100 μg/ml aIgG stimulated a 2.5-fold increase in histamine release that was statistically significant compared to spontaneous level ($p<0.05$). This effect is believed to be a result of FcgR cross-linking on the basophil cell surface membrane.

Addition to the cells of the nicotinic antagonist hexamethonium together with aIgG led to a decrease in the induced histamine release compared to both aIgG effect and "expected" effect calculated for combined IgG+Hex action (Figure 4, $p<0.05$). Expected estimate was calculated using corresponding values of separate effects of aIgG and hexamethonium (for these separate values, see Figure 3). As nicotinic antagonist hexamethonium significantly decreased the aggregated IgG basophil response, it can be concluded that autonomic cholinergic system of basophils, particularly nAChRs, takes part in the mechanisms of basophil degranulation and histamine secretion induced by aIgG through FcgRs. Nicotinic AchRs seem to play a costimulatory role in this process.

Moreover, the addition of muscarinic antagonist (at a dose of 0.17 mg/ml) to the above mixture of aIgG and hexamethonium resulted in further suppression of basophil response. Histamine release dropped to the background level. This effect of methacine was statistically significant compared to both aIgG effect and "expected" effect of IgG+Hex+Meth combination (Figure 4).

The effects of hexamethonium and methacine indicate that Ach signaling via both nicotinic and muscarinic AchRs are needed for and participate in the IgG stimulating effect on basophil histamine secretion. Cholinergic signaling seems to be evoked during FcgR cross-linking by IgG aggregates. It can include FcgR-activated enhancement of Ach synthesis with further autocrine self-stimulation of the cells that produced it or/and paracrine affecting the nearby cells.

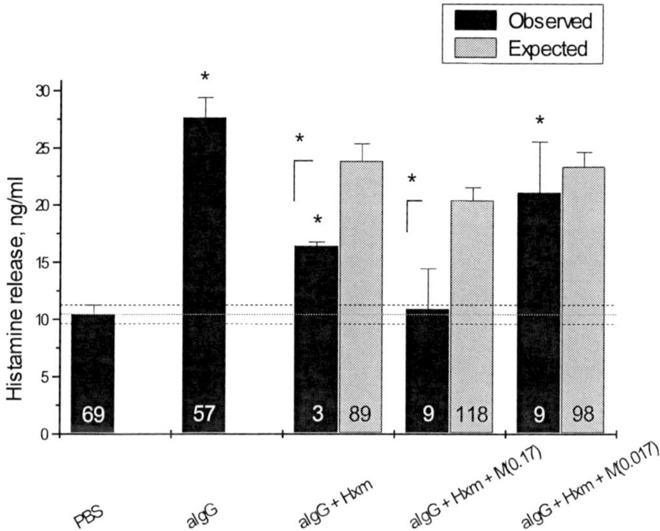

IgG – heat aggregated normal human IgG, 100 μg/ml; Hxm – hexamethonium, 4.16 mg/m; M(0.17) and M(0.017) – methacine, 0.17 and 0.017 mg/ml, respectively. Expected values were calculated using effect of aIgG and separate effects of Hxm, M(0.17) and M(0.017) that are shown on Figure 3. Numbers at the bottom of the columns represent the number of experiments.

Figure 4. Human blood basophil responses to aggregated IgG. Effect of nAChR and mAChR blockade.

The addition of a lower dose of muscarinic antagonist (0.017 mg/ml) to the aIgG plus hexamethonium mixture resulted in the cancellation of hexamethonium inhibitory effect. Really observed mean level of histamine release under the action of the aIgG+Hxm+Met(0,017) mixture does not differ not only from the calculated expected value but also from the observed effect of aIgG alone. This evidence supports the above suggestion that a low-dose effect of muscarinic antagonist methacine may be based on the inhibitory function of high affinity mAChRs occupied by small quantity of the drug molecules.

The results of this section show that aggregated IgG when affecting the cells involves more than one type of receptors. In addition to FcgRs which it binds and cross-links as their ligand it may concomitantly affect other receptor systems. One of such foreign receptor system that is obviously involved in the aIgG basophil response is cholinergic pair of two AchR types, nicotinic and muscarinic.

4. Basophil Responses to Human C-Reactive Protein and Cholinergic Drugs

As can be seen from figures 1 and 5, CRP activates histamine release from human blood basophils. After incubation of basophils with the combinations of CRP plus carbachol or CRP plus armine the significant inhibition of histamine release was registered compared with either the observed effect of CRP alone or with the calculated estimates of expected effects of the two agents of which each has demonstrated basophil activating activity. Instead of high levels of histamine release that we might expect based on the separate effects of CRP, carbachol and armine, we observe significantly lower histamine release in both cases. Really observed effects of CRP+Cch and CRP+armine significantly differed from both expected estimates (Figure 5, $p<0.05$ for both comparisons). This makes it possible to reject null hypothesis that there was no cooperation between CRP and carbachol as well as between CRP and armine. The data suggest that some kind of CRP interaction with these cholinergic agents does take place. As we have previously shown [Nazarov P.G. et al., 2007], CRP binds acetylcholine and masks its effects because of structural similarity between Ach and phosphocholine (PC, the major ligand of CRP).

The carbachol molecule is closely related to both Ach and PC molecules (Figure 6), which allows to assume that carbachol binding by CRP might occur. As to armine, the resulting product responsible for its biological activity is Ach as well. So, the results shown on Figure 5 suggest that low basophil responses to CRP combinations with carbachol and armine were associated with a decrease in local effective concentration of endogenous Ach (in the case of armine) or carbachol due to their binding by CRP rather than with an active mechanism of suppression.

Incubation of normal basophils with CRP combined with AchR antagonists, hexamethonium or methacine, resulted in significant enhancement of CRP histamine release-inducing activity. The levels of secreted histamine after stimulation of the cells with CRP+Hxm or CRP+M (with either dose of metacine, 0.17 or 0.017 μg/ml) were much higher than observed with CRP alone ($p<0.05$) or expected for CRP combination with these cholinergic antagonists (p<0.05 for any combination) (Figure 5).

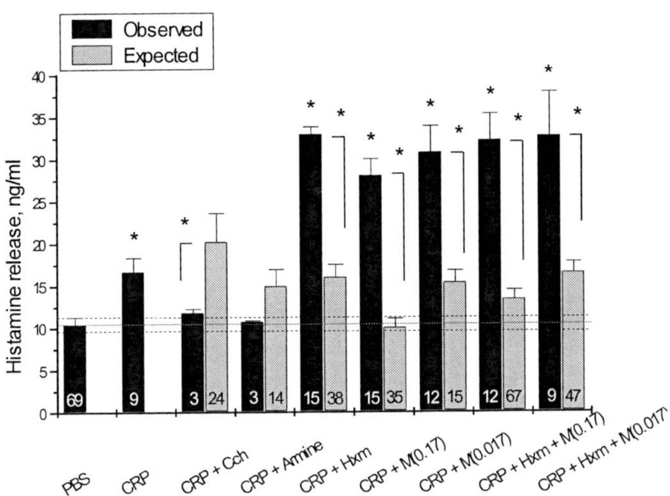

CRP – human CRP, 50 μg/ml; Cch – carbachol, μg/ml; armine – irreversible AchE inhibitor, 2 μg/ml; Hxm – hexamethonium, 4.16 mg/m; M(0.17) and M(0.017) – methacine, 0.17 and 0.017 mg/ml, respectively. Separate effects of Hxm, armine, M(0.17) and M(0.017) that were used in calculation of expected estimates plotted here are shown on Figure 3. Numbers at the bottom of the columns represent the number of experiments.

Figure 5. Human blood basophil responses to CRP: effect of nAChR and mAChR blockade.

In contrast to expected moderate decrease of CRP-induced basophil secretion in the presence of nAChR and mAChR blockers, both cholinergic antagonists enhanced basophil degranulation and histamine release. This situation is quite different from that showed above for aggregated IgG and its combinations with cholinergic blockers. Rather than lower the basophil reaction to CRP like observed with aIgG, AchR antagonists highly elevated it. It suggests that CRP can affect biological activity of hexamethonium and methacine by unknown mechanism.

We have reported earlier that CRP diminished the severity of the anaphylaxis reaction when applied either in the period of animal primary sensitization with a protein antigen or 30 min before the injection of a resolution dose of antigen to sensitized animals [Nezhinskaya G.I. et al., 2004, 2005]. Methacine in these experiments significantly reduced the severity of the anaphylactic shock. However, if methacine was injected together with CRP immediately before the resolution injection of the antigen, there was no decline in the intensity of the shock, which should be expected since methacine and CRP caused protective effect separately.

Instead a significant increase in the severity of shock was observed [Nezhinskaya G.I. et al., 2004, 2005]. This observation 110nfracted that CRP and methacine could interact and neutralize each other.

Structure	Compound
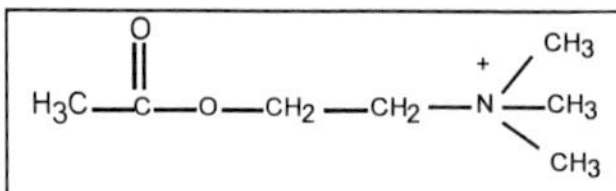 $H_3C-C(=O)-O-CH_2-CH_2-N^+(CH_3)_3$	Acetylcholine
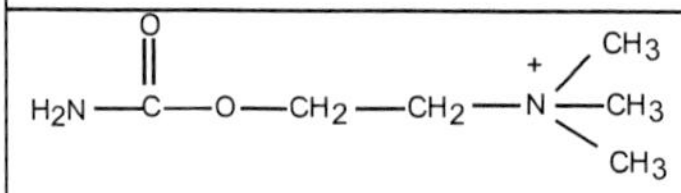 $H_2N-C(=O)-O-CH_2-CH_2-N^+(CH_3)_3$	Carbachol
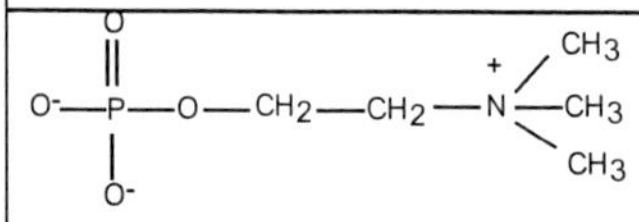 $O^--P(=O)(O^-)-O-CH_2-CH_2-N^+(CH_3)_3$	Phosphocholine
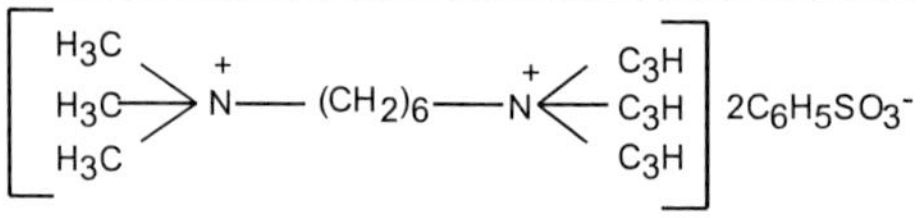 $[(H_3C)_3N^+-(CH_2)_6-N^+(C_3H)_3]\ 2C_6H_5SO_3^-$	Hexamethonium (1,6-bis-(N-trimethylammonium)-hexane dibenzolsuphonate)
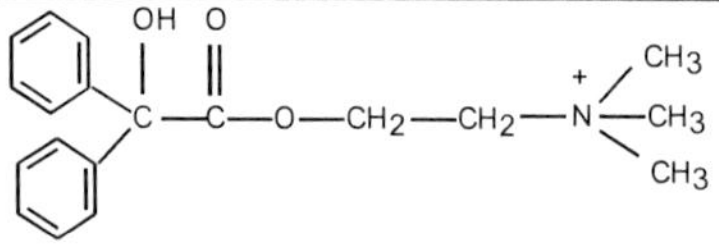 $(C_6H_5)_2C(OH)-C(=O)-O-CH_2-CH_2-N^+(CH_3)_3$	Methacine (beta-dimethylaminoethyl ether of benzyl acid, iodine methylate)
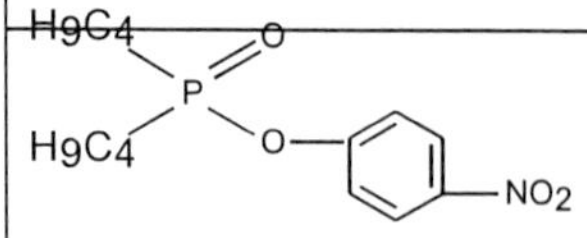 $(H_9C_4)_2P(=O)-O-C_6H_4-NO_2$	Armine (ethyl-p-nitrophenyl ether of ethylphosphonic acid)

Figure 6. The structure of cholinergic compounds used in the study.

Instead a significant increase in the severity of shock was observed [Nezhinskaya G.I. et al., 2004, 2005]. This observation 111nfracted that CRP and methacine could interact and neutralize each other.

The development of acute shock in those experiments was significantly accelerated by muscarinic acetylcholine receptor (mAChR) stimulation with aceclidine prior to antigen injection. So, enhanced acetylcholine (Ach) signaling through mAChRs caused the severity of shock to increase. The same result could be obtained with the nAChR blocker hexamethonium. Thus, an increase in shock severity can be provided by enhancement of muscarinic stimulation with either muscarinic agonist or nicotinic antagonist [Nezhinskaya G.I., Nazarov P.G., Evdokimova N.R., Losev N.A., Sapronov N.S., 2004].

Anaphylactic shock could be prevented with a combination of mAChR blocker methacine (applied 40 min prior) and AchE inhibitor neostigmine injected 15 min prior antigen.

Purified plasma proteins influenced methacine activity differently. Purified human CRP co-administered with methacine blocked effect of the latter whereas normal human IgG injection prevented shock [Nezhinskaya G.I., Nazarov P.G., Evdokimova N.R., Losev N.A., Sapronov N.S., 2004; Nezhinskaia G.I., Losev N.A., Nazarov P.G., Sapronov N.S., 2005].

As shown in Figure 6, hexamethonium and methacine contain similar backbones and identical trimethlammonium heads in their molecules. It makes them very similar to both phosphocholine and acetylcholine and leads to the suggestion that these drugs might be ligated by CRP as well. If so, CRP could affect their binding to appropriate AchRs and somehow modulate their signaling. The resulting data indicate that, whatever the mechanism of CRP interaction with these substances, the result of their joint action on basophils manifested in the form of enhancing the activity of the cells and more robust release of histamine.

Thus, the results of two studied FcgR-specific ligands, aggregated IgG and CRP, the reactant of acute phase of inflammation, have indicated common patterns of their influence on basophils, and features that distinguish them from each other. Common to aIgG and CRP is their activating effect on basophils. Both agents can activate normal basophils to the rapid release of histamine. Aggregated IgG is regarded as a model of immune complexes. Immune complexes are formed in many pathological and normal, physiological states. Our data suggest the ability of immune complexes to provide strong stimulus to basophils and possibly tissue mast cells.

Immune complexes containing IgG should evoke prompt release of vasoactive histamine and other proinflammatory mediators and cytokines produced by basophils. Consequently, in the sites of deposition of immune complexes (for example, in vessels impaired by atherogenesis, in kidney and joints affected with immune complex processes), basophils and mast cells should degranulate resulting in the release of their granules content and enhancement of its local impact on immunocompetent cells. It can be expected that the acute phase of inflammation, which is characterized by increasing the CRP gene expression and sharp strengthening of CRP protein production, will also be accompanied by significant increases in concentrations of histamine in the blood plasma and tissues due to the pentraxin activation of its release from the cells. It has been previously shown that CRP is synthesized not only by hepatocytes, but also by activated lymphocytes [Nazarov P.G., Sofronov B.N., 1983; Ikuta/Kuta, 1986]. Elevated local CRP production by activated lymphocytes and the related increased degranulation of basophils (mast cells) should take place in the sites of inflammation. According to the data presented herein, CRP and aggregated IgG show similar action on basophils in the absence of cholinergic drugs brought from outside. The difference between CRP and aggregated IgG is manifested only in artificial conditions, when blockers of Ach receptors are added to the cell system. If the impact of aggregated IgG on human basophils to take over standard, then the unusual effects of CRP on these cells can be explained by the possessing a ligand-binding activity by the pentraxin and its ability of binding the cholinergic antagonists due to their similarity to its ligands. The situations where CRP can meet such substances, are possible in the clinic, with medicines that contain in its structure quaternary ammonium base. In these cases, if there is an acute inflammation (and high CRP in the blood), perverse reactions may develop to such drugs because of their interaction with CRP and the CRP impact on their signaling (and therapeutic) functions.

5. Joint Effects of CRP with aIigG and Antibody to CD16

The data presented on Figure 7 shows the impact on the basophil secretory activity of three ligands, reacting with Fc-gamma receptors: aggregated IgG, C-reactive protein and monoclonal antibody to the Fc-gamma-type receptor III (low affinity). All three agents cause significant strengthening of histamine secretion (for all cases $p < 0.05$), of which IgG – most expressed.

With the incubation of the cells with two activators at once, CRP and aIgG, there was no addition of their effects. Instead, there was a decrease in histamine secretion compared not only with the expected mean of their separate effects (not shown), but also with the effect of any of these agents.

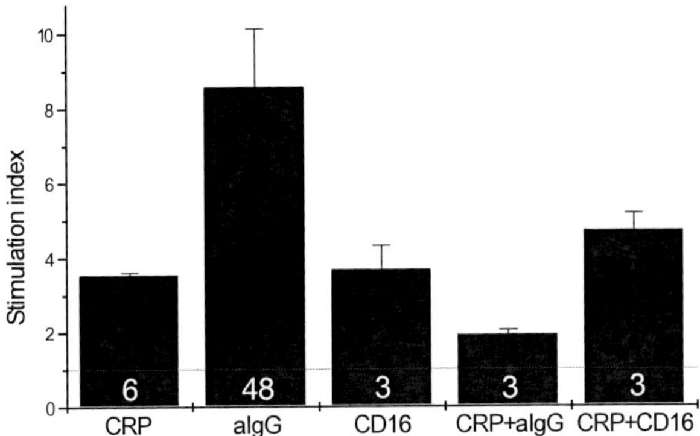

CRP – human CRP, 50 μg/ml; aIgG – 100 μg/ml, mouse anti-human CD16 – 20 μl/ml. Numbers at the bottom of the columns represent the number of experiments. Means ± SEM are plotted. The red control line represents the mean spontaneous histamine release in the presence of PBS taken to be 1.0.

Figure 7. Human blood basophil responses to CRP, aIgG and anti-CD16, and their combinations.

This may indicate that CRP and aIgG interact with the same receptors on the surface of basophils and compete with each other for cell surface binding sites or for intracellular messengers.

Some data indicates that CRP interacts with almost all types of FcgRs, including FcgRIIb. It has been reported that CRP binds leukocyte FcgRIIa (CD32) [Bharadwaj D. et al., 1999]. CRP induces ICAM-1, VCAM 1 and IL-8 up-regulation and down-regulates endothelial NO synthase (eNOS) via CD32 and CD64 and through NF-kappaB-mediated pathway [Yao-Jen Liang et al., 2006; Devaraj S. et al., 2005]. CRP up-regulates monocyte endothelial adhesion by activation of NF-kappaB through engaging the FcgRs CD32 and CD64, but not CD16. Biotinylated CRP bound to cells was colocalized with CD32 and CD64, but not with CD16 [Devaraj S., Davis B., Simon S.I. et al., 2006]. Also, preincubation with anti-CD32 and CD64 antibodies significantly inhibited binding of CRP to human aorta endothelial cells whereas antibodies to CD16 had no effect [Devaraj S., 2005]. FcgRIIs are thought to be the principal high affinity receptors for CRP expressed in endothelium. CRP prevents eNOS activation in cultured endothelium by a proper agents due to the interaction with FcgRIIb and inhibition of PP2A. In cells not expressing FcgRIIb, CRP did not antagonize eNOS activation while in cells expressing FcgRIIb, CRP blunted eNOS activation, indicating that the action of CRP requires FcgRIIb [Mineo C. et al., 2005].

Among FcgRs, the FcgRIIb contains an immunoreceptor tyrosine-based inhibitory motif (ITIM) in its intracellular part, which can inhibit activation after nfractedi with ITAM containing FcgRs (all FcgRs except FcgRIIb). In humans, FcgRIIa is an activation receptor, while FcgRIIb is an inhibitory receptor [Daeron M., 1997; Ravetch J.V., Bolland S., 2001].The existence of ITIM-containing FcgRIIb for IgG has been shown on mast cells and basophils as well [Katz H.R., 2002]. Cross-linking of FcgRIIb with FceRI leads to inhibition of basophil triggering [Wiggington S.J. et al., 2008] and can inhibit the release of allergic mediators characteristic of type I hypersensitivity reactions.

Thus, FcgRIIb could be implicated in modulating human basophil secretory activity. Activation of FcgRIIb could occur under the joint action on basophils of aIgG and CRP as well as of aIgG and AchR antagonists (see Section 3, figure 4). In the latter case the activation of this suppressive receptor subtype may occur as a result of its 114nfracted114i with nicotinic or muscarinic AchRs.

In contrast, basophil response to the joint action of CRP and anti-CD16 antibody did not show any inhibition. Instead, CRP and anti-CD16 effects were partially summarized. This result favors the view that CRP does not interact with CD16, and thus low affinity FcgRIIIs are not the main binding sites of this pentraxin on the blood basophils. The cross-linking of FcgRs by CRP on the one hand and CD16 molecules by antibodies on the other seems to occur independently and activates independent signaling mechanisms leading to basophil activation. So, under simultaneous stimulation of the cells with these two activators (e.g. CRP and anti-CD16) additive stimulatory effect was observed.

6. Normal Human Blood Basophil Responses to Repeated Activation in Vitro

Currently, there is little data on the survival of normal blood basophils in culture and ability to resynthesize their mediators *de novo* [Dvorak A.M. et al., 1982, 1985]. Information about normal basophil recovery/regranulation in vitro under activation through FcgRs by ligands like aggregated IgG and CRP we could not find in the literature. The study of this issue was the aim of our work. We have compared the survival of basophils during the cultivation of leukocytes in vitro, and investigated the ability of human blood basophils to degranulate repeatedly with re-release of histamine *in vitro*.

As our experiments showed, healthy donor blood basophils remain alive in leukocyte culture and able to produce histamine even 2-3 days after the beginning of cultivation *in vitro*.

The first stimulation of basophils was applied immediately after the start of cultures and lasted for 40 min. The cultures were stimulated with normal human heat aggregated IgG (50 or 500 μg/ml), human CRP (50 μg/ml), concanavalin A (10 or 50 μg/ml) or carbachol (1 μg/ml). Control cultures received PBS. After 40 min at 37°C supernatants were removed for histamine determination. This first basophil activation resulted in significant histamine release in response to aIgG, ConA, CRP and carbachol. The most active stimulant was carbachol (Figure 8). After that activated and control cells were washed thrice with PBS under centrifugation at 300 g for 10 min, and the last change of washing fluid was collected for histamine control. As was determined, it contained no measurable quantity of histamine. Then the cells were resuspended in fresh medium and cultivated for further 24 hr without stimulants. After 24 hours the activation of the cultures was repeated by adding the same stimulants. The cells were incubated with them for 40 min and supernatants were again collected to measure histamine release. Results of the first and second (at 24 hr) activation of basophils are shown in Figure 8.

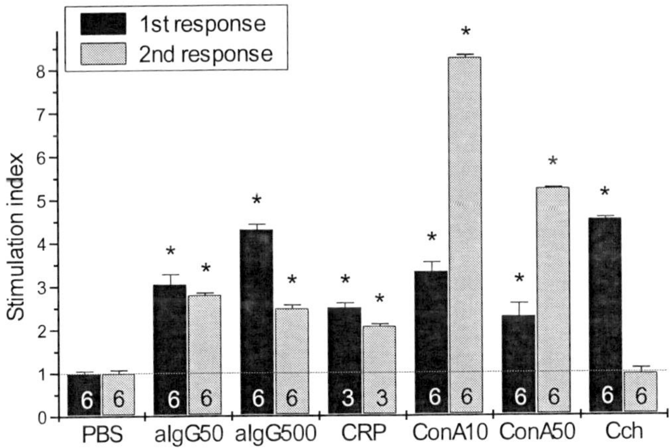

Means and standard errors of mean are shown. Normal human aggregated IgG was used in two doses – 50 and 500 μg/ml, CRP in a dose of 50 μg/ml, concanavalin A – in doses of 10 and 50 μg/ml, carbachol – 1 μg/ml. Numbers at the bottom of the columns represent the number of cultures. Asterisk – significantly different from the effect of PBS ($p<0.05$ or less).

Figure 8. Normal human blood basophil responses to repeated activation by aggregated IgG, CRP and concanavalin A in vitro.

As can be seen from Figure 8, after 24 hours of the start of cultivation the spontaneous basophil degranulation increased and histamine release was elevated. Significant histamine release at this term was induced by both doses of aIgG, CRP and ConA. The response to carbachol was not observed.

There are published data, which show that basophils in culture can live for a week. The authors of this observation have cultured cells in the presence of recombinant IL-3, a factor known to support basophil differentiation. According to their suggestion, IL-3 ensured basophils survival in vitro during that period [Yamaguchi M. et al., 1996]. However, the role of IL-3 in basophil survival is hard to judge based on the reported data, because the authors have not showed the data on basophil survival in cultures without adding IL-3. Unlike the work of Yamaguchi, we cultured cells without adding IL-3.

Carbachol is a nonmetabolizable analog of Ach, which possesses the same as Ach signaling activity, but differs from the Ach so that is not subjected to splitting enzymes. Carbachol is commonly used in experiments in vitro as a convenient substitute for Ach [Jensen A.A. et al., 2003].

We have obtained data that confirm the fact that basophils like mast cells [Falcone F.H. et al., 2006] are sensitive to cholinergic mediators and respond to carbachol with increased secretion of biologically active substances. Leukocytes responded *in vitro* to the first incubation with the carbachol with high elevation of histamine release. However, on the second day of cultivation, the cells showed no response to carbachol, although did respond significantly to the repeated stimulation with other activators (aggregated IgG, CRP and ConA).

Perhaps in this case the known phenomenon of desensitization was observed that is the loss of receptor sensitivity at re-exposition to the agonist [Picciotto M.R. et al., 2008]. Two types of cells desensitization has been described: homologous – when the cells no longer respond to the repeated impact of the same activator, which was used first time, and heterologous – when the cells do not respond to any reactivating stimulant. In our experiments, homologous desensitization to carbachol was observed: basophils that degranulated in response to the first contact with this carbachol, did not respond to reactivation with this agent.

At the same time, there was no distinct loss of basophil reactivity to repeated activation with plant lectin ConA, aggregated IgG and CRP. This indicates that the mechanisms of desensitization, described for neurotransmitters and their receptors, may not be used by other receptors and their signaling pathways (in our case mannose containing receptors of ConA and FcgRs). Alternatively, desensitization mechanisms could be compensated with some biochemical factors that do not affect the signaling of Ach receptors.

Thus, it can be concluded that basophils survive in culture for at least 24 hours and retain the ability to restore histamine and release it repeatedly in response to reactivation.

Basophis completed degranulation and histamine release upon stimulation with FcgR-specific ligands such as aggregated IgG and CRP or mannose-specific ligand ConA can restore their histamine content overnight to secrete it again at restimulation with the same factors. After cholinergic activation of basophils by carbachol the desensitization develops, which manifests itself in the loss of the cell ability to respond to the repeated exposure to this activator.

ACKNOWLEDGMENT

The study was supported by Russian Foundation of Basic Research, grant 06-04-49629.

REFERENCES

Beekman, JM; Bakema, JE; van de Winkel, JG; Leusen, JH. Direct interaction between FcgammaRI (CD64) and periplakin controls receptor endocytosis and ligand binding capacity. *Proc. Natl. Acad. Sci. U.S.A.* 2004, 101(28), 10392-10397.

Bharadwaj, D; Mold, C; Markham, E; Du Clos, TW. Serum amyloid P component binds to Fcg receptors and opsonizes particles for phagocytosis. *J. Immunol.*, 2001, 166(11), 6735–6741.

Bharadwaj, D; Stein, MP; Volzer, M; Mold, C; Du Clos, TW. The major receptor for C-reactive protein on leukocytes is Fcgamma receptor II. *J. Exp. Med.*, 1999, 190(4), 585-590.

Bühring, HJ; Streble, A; Valent, P. The basophil-specific ectoenzyme E-NPP3 (CD203c) as a marker for cell activation and allergy diagnosis. *Int. Arch. Allergy Immunol.*, 2004, 133(4), 317-329.

Devaraj, S; Davis, B; Simon, SI; Jialal, I. CRP promotes monocyte-endothelial cell adhesion via Fc-gamma receptors in human aortic endothelial cells under static and shear flow conditions. *Am. J Physiol. Heart Circ. Physiol.*, 2006, 291(3), H1170–H1176.

Devaraj, S; Du Clos, TW; Jialal, I. Binding and internalization of C-reactive protein by Fcgamma receptors on human aortic endothelial cells mediates biological effects. *Arterioscler. Thromb. Vasc. Biol.*, 2005, 26(7), 1359-1363.

Du Clos, TW; Mold, C. C-reactive protein an activator of innate immunity and a modulator of adaptive immunity. *Imm. Res.*, 2004, 30(3), 261-277.

Du Clos, TW; Mold, C. The role of C-reactive protein in the resolution of bacterial infection. *Curr. Opin. Inf. Dis.*, 2001, 14(3), 289-293.

Dvorak, AM; Colvin, RB; Monahan, RA. Immunoferritin electron microscopic studies with antibasophil serum of guinea pig basophil degranulation and regranulation in vitro. *Clin. Immunol. Immunopathol.*, 1985, 37(1), 63-76.

Dvorak, AM; Galli, SJ; Morgan, E; Galli, AS; Hammond, ME; Dvorak, HF. Anaphylactic degranulation of guinea pig basophilic leukocytes. II. Evidence for regranulation of mature basophils during recovery from degranulation in vitro. *Lab. Invest.*, 1982, 46(5), 461-475.

Falcone, FH; Zillikens, D; Gibbs, BF. The 21st century renaissance of the basophil? Current insights into its role in allergic responses and innate immunity. *Exp. Dermatol.*, 2006, 15(11), 855-864.

Fujimoto, T; Sato, Y; Sasaki, N; Teshima, R; Hanaoka, K; Kitani, S. The canine mast cell activation via CRP. *Biochem. Biophys. Res. Commun.*, 2003, 301(1), 212–217.

Grützkau, A; Smorodchenko, A; Lippert, U; Kirchhof, L; Artuc, M; Henz, BM. LAMP-1 and LAMP-2, but not LAMP-3, are reliable markers for activation-induced secretion of human mast cells. *Cytometry A.*, 2004, 61(1), 62-68.

Gupta, N; Scharenberg, AM; Fruman, DA; Cantley, LC; Kinet, JP; Long, EO. The SH2 domain-containing inositol 5-phosphatase (SHIP) recruits the p85 subunit of phosphoinositide 3-kinase during FcgRIIb1-mediated Inhibition of B cell receptor signaling. *J. Biol. Chem.*, 1999, 274(11), 7489-7494.

Horowitz, J; Volanakis, JE; Briles, DE. Blood clearance of Streptococcus pneumoniae by C-reactive protein. *J. Immunol.*, 1987, 138(8), 2598-2603.

Ikuta, T; Okubo, H; Ishibashi, H; Okumura, Y; Hayashida, K. Human lymphocytes synthesize C-reactive protein. *Inflammation.*, 1986, 10(3), 223-232.

Jensen, AA; Mikkelsen, I; Frølund, B; Bräuner-Osborne, H; Falch, E; Krogsgaard-Larsen, P. Carbamoylcholine homologs: novel and potent agonists at neuronal nicotinic acetylcholine receptors. *Mol. Pharmacol.*, 2003, 64(4), 865-875.

Katz, HR. Inhibitory receptors and allergy. *Curr. Opin. Immun.*, 2002, 14(6), 698-704.

Kawashima, K; Fujii, T. Extraneuronal cholinergic system in lymphocytes. *Pharmacol. Ther.*, 2000, 86(1), 29-48.

Kirkpatrick, CJ; Bittinger, F; Nozadze, K; Wessler, I. Expression and function of the non-neuronal cholinergic system in endothelial cells. *Life Sci.*, 2003, 72(18-19), 2111-2116.

Knol, EF; Mul, FP; Jansen, H; Calafat, J; Roos, D. Monitoring human basophil activation via CD63 monoclonal antibody 435. *J. Allergy Clin. Immunol.*, 1991, 88(3 Pt 1), 328-338.

Lange, A; Karabon, L; Tomeczko, J. IL-6- and IL-4-related proteins (C-reactive protein and IgE) are prognostic factors of asbestos-related cancer. *Ann. N.Y. Acad. Sci.*, 1995, 762, 435-438.

Liang, YJ; Shyu, KG; Wang, BW; Lai, LP. C-reactive protein activates the nuclear factor-κB pathway and induces vascular cell adhesion molecule-1 expression through CD32 in human umbilical vein endothelial cells and aortic endothelial cells. *J. Mol. Cell.*, Cardiol. 2006, 40(3), 412-420.

Marnell, LL; Mold, C; Volzer, MA; Burlingame, RW; Du Clos, TW. C-reactive protein binds to Fc gamma RI in transfected COS cells. *J. Immunol.*, 1995, 155(4), 2185-2193.

Metzelaar, MJ; Wijngaard, PL; Peters, PJ; Sixma, JJ; Nieuwenhuis, HK; Clevers, HC. CD63 antigen. A novel lysosomal membrane glycoprotein, cloned by a screening procedure for intracellular antigens in eukaryotic cells. *J. Biol. Chem.*, 1991, 266(5), 3239-3235.

Mineo, C; Gormley, AK; Yuhanna, IS; Osborne-Lawrence, S; Gibson, LL; Hahner, L; Shohet, RV; Black, S; Salmon, JF; Samols, D; Karp, DR; Thomas, GD; Shaul, PW. FcgammaRIIB mediates C-reactive protein inhibition of endothelial NO synthase. *Circ. Res.*, 2005, 97(11), 1124-1131.

Mohamed, HA; Mosier, DR; Zou, LL; Siklós, L; Alexianu, ME; Engelhardt, JI; Beers, DR; Le, WD; Appel, SH. Immunoglobulin Fc gamma receptor promotes immunoglobulin uptake, immunoglobulin-mediated calcium increase, and neurotransmitter release in motor neurons. *J. Neurosci. Res.*, 2002, 69(1), 110-116.

Mold, C; Gresham, HD; Du Clos, TW. Serum amyloid P component and C-reactive protein mediate phagocytosis through murine FcgRs. *J. Immunol.*, 2001, 166(2), 1200–1205.

Nakayama, S; Du Clos, TW; Gewurz, H; Mold, C. Inhibition of antibody responses to phosphocholine by C-reactive protein. *J. Immunol.*, 1984, 132(3), 1336–1340.

Nazarov, PG. Reactants of Acute Phase of Inflammation. 2001, Nauka, St. Petersburg. 423 (In Russian).

Nazarov, PG; Krylova, IB; Evdokimova, NR; Nezhinskaya, GI; Butyugov, AA. C reactive protein: a factor of inflammation binding and inactivating acetylcholine. *Cytokines and Inflammation.*, 2006, 6(4), 32-35. (in Russian).

Nazarov, PG; Krylova, IB; Evdokimova, NR; Nezhinskaya, GI; Butyugov, AA. C-reactive protein: a pentraxin with anti-acetylcholine activity. *Life Sci.*, 2007, 80(24-25), 2337–2341.

Nazarov, PG; Nezhinskaya, GI; Krylova, IB. Pentraxins: antiarrhythmogenic factors of inflammation. *Ann. Arrhythmol.*, 2005, 2(suppl), 20.

Nazarov, PG; Sofronov, BN. Synthesis of C-reactive protein by lymphoid cells. In: Mediators of Immune Response in Experiments and Clinics. Institute of Immunology, USSR Ministry of Health, Moscow, 1983, 112–113. (In Russian).

Nezhinskaya, GI; Losev, NA; Nazarov, PG; S apronov, NS. Effect of acetylcholine and C-reactive protein on regulation of anaphylactic shock in guinea pigs. *Eksp. Klin. Farmakol.*, 2005, 68(4), 49-52. (In Russian)

Nezhinskaya, GI; Nazarov, PG; Evdokimova, NR; Losev, NA; Sapronov, NS. Cholinergic regulation of anaphylactic shock: impact of C-reactive protein. *Cytokines and Inflammation.*, 2004, 3(1), 39-41. (in Russian).

Nieuwenhuis, HK; van Oosterhout, JJ; Rozemuller, E; van Iwaarden, F; Sixma, JJ. Studies with a monoclonal antibody against activated platelets: evidence that a secreted 53,000-molecular weight lysosome-like granule protein is exposed on the surface of activated platelets in the circulation. *Blood.*, 1987, 70(3), 838-845.

Oishi, K; Yoshizumi, T; Seki, N; Uchida, MK. Carbachol-induced membrane ruffling and Its desensitization in rat basophilic leukemia (RBL-2H3) cells transfected with m3 muscarinic acetylcholine receptors. *Biol. Pharm. Bull.*, 2004, 27(10), 1544-1548.

Picciotto, MR; Addy, NA; Mineur, YS; Brunzell, DH. It is not "either/or": Activation and desensitization of nicotinic acetylcholine receptors both contribute to behaviors related to nicotine addiction and mood. *Prog. Neurobiol.*, 2008, 84(4), 329-342.

Ravetch, JV; Bolland, S. IgG Fc receptors. *Annu Rev Immunol.*, 2001, 19, 275-290.

Suresh, MV; Singh, SK; Agrawal, A. Interaction of calcium-bound C-reactive protein with fibronectin is controlled by pH: in vivo implications. *J. Biol. Chem.*, 2004, 279(50), 52552-52557.

Suzuki, K; Kitani, S; Takaishi, T; Ito, K; Ra, C; Morita, Y. Nonreleasing basophils convert to releasing basophils by culturing with IL-3. *J. Allergy Clin. Immunol.*, 1996, 97(6), 1279-1287.

Szalai, AJ; Briles, DE; Volanakis, JE. Human C-reactive protein is protective against fatal Streptococcus pneumoniae infection in transgenic mice. *J. Immunol.*, 1995, 155(8), 2557-2563.

Tseng, J; Mortensen, RF. Binding of human C-reactive protein (CRP) to plasma fibronectin occurs via the phosphorylcholine-binding site. *Mol. Immunol.*, 1988, 25(8), 679-686.

Tseng, J; Mortensen, RF. The effect of human C-reactive protein on the cell-attachment activity of fibronectin and laminin. *Exp. Cell Res.*, 1989, 180(2), 303-313.

Van de Winkel, JG; Capel, PJ. Human IgG Fc receptor heterogeneity: molecular aspects and clinical implications. *Immunol. Today.*, 1993, 14(5), 215-221

Van de Winkel, JGJ; Anderson, CL. Biology of human immunoglobulin G Fc receptors. *J. Leukoc. Biol.*, 1991, 49(5), 511–524.

Wessler, I; Kilbinger, H; Bittinger, F; Unger, R; Kirkpatrick, CJ. The non-neuronal cholinergic system in humans: expression, function and pathophysiology. *Life Sci.*, 2003, 72(18-19), 2055-2061.

Wigginton, SJ; Furtado, PB; Armour, KL; Clark, MR; Robins, A; Emara, M; Ghaemmaghami, AM; Sewell, HF; Shakib, F. An immunoglobulin E-reactive chimeric human immunoglobulin G1 anti-idiotype inhibits basophil degranulation through cross-linking of FcepsilonRI with FcgammaRIIb. *Clin. Exp. Allergy.*, 2008, 38(2), 313-319.

Xiao, Y; Kellar, K J. The comparative pharmacology and up-regulation of rat neuronal nicotinic receptor subtype binding sites stably expressed in transfected mammalian cells. *J. Pharmacol. Exp. Ther.*, 2004, 310(1), 98-107.

Zhang, D; Sun, M; Samols, D; Kushner, I. STAT3 participates in transcriptional activation of the C-reactive protein gene by interleukin-6. *J. Biol. Chem.*, 1996, 271(16), 9503-9509.

INDEX

A

B

C

D

E

J

K

L

M

N